NEO-CLASSICAL
ART IN HOTELS
新古典酒店艺术

任绍辉／编　　于 芳　常文心／译

辽宁科学技术出版社

NEO-CLASSICAL
ART IN HOTELS
新古典酒店艺术

任绍辉 / 编　　于 芳 常文心 / 译

辽宁科学技术出版社

Preface 前言

Once upon a time in the great hotels luxury meant golden taps and marble walls, materials that are banned by contemporary design and are replaced with simplicity and purity.

These words today have a new meaning. It is ironic how the luxury industry has destroyed individualism; the most famous shopping strips around the world feature the same brands, the same products, the same hotels.

Gone are the days when arriving in New York or Paris matched with the excitement of finding unusual and extraordinary spaces. The collection in this book represents hotels that rewrite the spaces with the languages of creativity. The image captures the true essence of habitat and gives back to the guest the seduction coming from the materials.

Balancing contemporary taste and respect for the past, an old farm or a manor house becomes a welcoming place, a luxury resort or a small charm hotel. The guests are pampered and surrounded by unique and precious elements.

This is my concept of luxury and elegance.

It is time for original, vibrant, magical and enlightened visionaries. One of the most exciting parts of a designer's job is to treat some areas of the hotel as a stage. If you love theatre and its most spectacular angles... let the show begin!

The secret of our lifestyle in the near future is illustrated by the trend site we are visiting. The atmosphere depends not only on the area the designers are working on but also on which ways we can transform it. No wonder the most passionate aspect of our mission is to welcome people in our space and make sure they feel good in it. The interior design's goal is to create a setting for the life we want to live. It should make us feel better; it's the interior design that creates it. As it is in theatre, light is the key that can modify the atmosphere more than any other component. The choice of furnishing is just as important because of the imprinting they leave on the space. There is a fine line between perfection and banality; good hotels rely on dimensions and proportions. The mixture of humble and luxurious elements can create an environment that escapes classification. A good design is not only a matter of colours, floors or accessories. A good designer works on the projects from the beginning, striking the right balance between space and functionality with the purpose of creating areas that are both comfortable and balanced.

Alvin Grassi, designer in Italy

从前，奢华酒店意味着黄金水龙头和大理石墙面。这些材料为现代设计所摒弃，并早已被简约、纯粹的风格所替代。

如今，这些词语被赋予了全新的意义。讽刺的是，奢侈品产业已经摧毁了个人主义；全世界著名的商业街区全都充斥着同样的品牌，同样的产品，以及同样的酒店。

在纽约或者巴黎寻找奇妙空间的乐趣的日子早已经一去不返。本书所精选的酒店重新以创意语言诠释了空间。酒店捕捉了居住地的真谛，重新让宾客感受到了来自材料的诱惑。

在现代品位和历史中获得平衡，旧农场或老庄园也能成为备受欢迎的景点、奢华的度假村或小型魅力酒店。宾客们将得到无微不至的照顾，被独特而珍贵的元素所包围。

这就是我对奢华和优雅的理解。

这个时代充满了独创、活力、神奇而进步的愿景。设计师工作中最令人兴奋的一部分就是将酒店的某些区域作为一个舞台来进行设计。如果你热爱剧院和它的华美，你就一定会喜欢这些酒店。

我们所造访的潮流场所描绘了我们未来的生活风格。气氛的营造不仅取决于场地本身的特质，还在于设计师对它的改造方式。设计师工作中最富激情的一部分就是迎接人们来到我们设计的空间并保证他们感到舒适自在。室内设计的终极目标就是打造一个我们所渴望的生活场景。它应当通过室内设计赋予我们更好的体验。

正如在剧院中一样，酒店的灯光是最能调节气氛的元素。装饰陈设的选择也同等重要，因为它们为空间留下了印记。完美和平凡仅有一线之差，酒店应以恰当的规模和比例而取胜。低调与奢华元素的结合能够营造出超凡脱俗的环境。好设计不仅取决于色彩、地面或装饰品。一位好的设计师应当在空间与功能之间实现良好的平衡，打造舒适而和谐的空间环境。

意大利设计师　阿尔文·格拉西

CONTENTS 目录

008　Chapter One　Overview for Neo-classical Hotels
第一章 新古典酒店概述

010　Origin of Neo-classicism
新古典主义风格的起源

011　Characteristics of Neo-classicism
新古典主义风格的特点

017　Manifestations of Neo-classicism in Different Countries
新古典主义风格在各个国家的表现形式

027　Characteristics of Neo-classical Hotels
新古典主义风格酒店的设计特点

034　Chapter Two　Spatial Layout of Neo-classical Hotels
第二章 新古典酒店的空间分布

038　Reception Area
接待区

　　　The Fairmont Palliser
　　　帕里瑟尔费尔蒙特酒店

048　Accommodation Area
住宿区

　　　Beau-Rivage Palace
　　　美岸大酒店

058　Dining Area
餐饮区

　　　Taleon Imperial Hotel
　　　塔里昂帝国酒店

066　Public Area
公共活动区

　　　Eynsham Hall
　　　艾恩汉姆会馆酒店

078　Chapter Three　Interface Design in Neo-classical Hotels
第三章 新古典酒店的界面设计

084　Ceiling Design
顶面设计

　　　Grand Hotel Kronenhof
　　　科隆霍夫大酒店

094　Wall Design
墙面设计

　　　Villa Le Rose
　　　玫瑰别墅酒店

106　Floor Design
地面设计

　　　The Dorchester
　　　多尔切斯特酒店

114　Chapter Four　Furniture in Neo-classical Hotels
第四章 新古典酒店的家具

120　Chairs
椅类家具

　　Chateau Mcely
　　梅斯丽城堡酒店

128　Tables
桌类家具

　　The Langham Huntington Hotel
　　亨廷顿朗廷酒店

138　Cabinets
柜类家具

　　Grand Hotel Vesuvio
　　维苏威大酒店

144　Chapter Five　Classic Decorations of Neo-classical Hotels
第五章 新古典酒店的经典配饰

150　Order
柱式

　　Fairmont Grand Hotel Kyiv
　　基辅费尔蒙特大酒店

158　Fireplace
壁炉

　　Bovey Castle
　　波维城堡酒店

168　Crystal Lamp
水晶灯

　　Charleston Place
　　查尔斯顿酒店

176　Textile
织物

　　The Milestone Hotel and Apartments
　　里程碑酒店

190　Chapter Six　Modernness and Tradition in the Neo-classical Hotel Design
第六章 新古典酒店设计中的现代与传统

196　Coordination between Traditional Designs and Modern Facilities
传统设计与现代设施的协调

　　Hôtel Beau-Rivage, Genève
　　日内瓦美岸酒店

208　Application of Modern Materials in Traditional Environment
现代材料在传统环境中的应用

　　Brenners Park-Hotel & Spa
　　布莱纳斯公园酒店

218　Renovation of Historical Sites
历史遗迹的修复

　　Fairmont Le Montreux Palace
　　费尔蒙特蒙特勒皇宫酒店

230　**INDEX** 索引

Chapter One
OVERVIEW FOR NEO-CLASSICAL HOTELS
新古典酒店概述

Chapter One
OVERVIEW FOR NEO-CLASSICAL HOTELS

1. Origin of Neo-classicism

The neo-classicism has its origin in 18th century, where the word "classicism" indicates that the style inherits the essence from Ancient Roma and Ancient Greece style, and the word "neo" differentiates the style from the style in classical and the Renaissance period. In the 18th century, rococo art prevailed in Europe. Delicate, complicated and crafted rococo art incisively and vividly presented extravagance and ostentation in French palace, especially in the interior design. Soft palette and carvings were adopted in abundance, and carvings with flowers and plants patterns added intense female colour. However, the palace style gradually declined in the wake of a series of drastic evolutions in the European continent of the 18th century (Figure 1). The 18th century was a period full of tremendous social transformation and ideological revolution. Academically, as a result of renaissance, the enlightenment movement fighting against feudal government and Church's ideology was launched amid fire and thunder on ideological field; technologically, the development of commerce facilitated the birth of numerous new technologies, and then opened a gate into the industrial revolution and achieved a transition from handicraft industry to mechanical industry; politically, as War of Independence broke out in America, as well as, French revolution erupted in France, the democratic ethos was born. These changes influenced the development of culture and art stealthily, and a trend of rational science replaced the prerogative of church and royal, which made art more simple and civilian-oriented. During the latter half of 18th century, rococo style as a transition was on the wane in the field of architectural and interior design, and designers and artists started searching for a simpler and more real style. The excavations to Pompeii and Herculaneum in Italy led the focus towards Ancient Greece and Ancient Rome, and then a series of books concerning architectural details were published, which let more people to comprehend the construction measures of classical architectures. Precise geometric proportions,

1. 新古典主义风格的起源

新古典主义风格起源于18世纪，"古典"指出了这种风格对古罗马和古希腊风格的继承，而"新"则又把它和古典时期以及文艺复兴时期的风格做出了区别。18世纪的欧洲是洛可可艺术风行的时代，纤巧繁复精雕细琢的洛可可艺术风格淋漓尽致地表现了法国宫廷的奢靡浮华，尤其是在室内装饰上，柔和的色泽、曲线的大量使用、花草纹样的雕刻为室内增添了浓重的女性色彩，但这种"宫廷风格"却随着18世纪欧洲大陆发生的一系列重大变化逐渐没落（图1）。18世

Figure 1. Rococo church is designed as total works of art with reliefs and wall paintings.

图1 洛可可风格教堂带有精美的浮雕和壁画。

Figure 2. The Pantheon is a building in the Latin Quarter in Paris. It was originally built as a church dedicated to St. Genevieve, after many changes, now functions as a secular mausoleum containing the remains of distinguished French citizens.

图2 先贤祠位于巴黎拉丁区，最初是一座名叫圣·热内维耶瓦（St. Genevieve）的教堂，几经周折后，现在被用来安葬法国具有突出贡献人的遗体。

ordered symmetry and magnificent and simple expression satisfied the pursuit to pure and real art. In a new social context, the form was granted a new connotation, and becames the unique neo-classicism.

2. Characteristics of Neo-classicism
2.1 Architectural Characteristics

The elements in neo-classicism origins from classical architectures, and therefore the elements in classical architectures are embodied clearly. Symmetry and balance are the main characteristics in neo-classicism, which are exactly the proper proportion that architectures of ancient Greece and Rome emphasis. The most distinct component forming the neo-classical architectures is the use of order. On the exterior, order can balance the proportion of architecture. At the same time, it is also the symbol of architecture in the period of ancient Greece and Rome. The reasonable use of order not only can make the building more spectacular apparently, but also can provide the powerful support and the sense of balance. Pantheon in Paris (Figure 2) can be called the paragon among the buildings in the early period of neo-classicism.

纪是欧洲社会大转型、思想大变革的时代，在学术上，由于文艺复兴的推动，思想领域开展了一场反对封建统治和教会思想的轰轰烈烈的启蒙运动；在科技上，商业的发展催生出了众多新技术的发明，从而开启了工业革命的大门，完成了从手工业向机器大工业的过渡；政治上，美国爆发了独立战争，法国爆发了法国大革命，产生了民主的思潮；这些变化都悄悄地影响着文化艺术的发展，理性科学的思想逐渐取代了教会和皇室的特权，艺术的形式也更加质朴和平民化。18世纪后半叶，在建筑和室内设计领域，过度装饰的洛可可风格开始逐渐衰落，设计师和艺术家们开始寻找更加朴实和真实的风格，而意大利庞贝古城和赫库兰尼姆古城的发掘把人们的视线引入到了古希腊和古罗马时代，随后一连串有关古典建筑细部书籍的陆续出版使古典建筑的设计建造方法得到了更多人的了解，古典建筑严谨的几何比例、有序的对称关系和宏伟质朴的表达方式满足了人们对纯粹真实艺术的追求，在新的社会背景下，这种形式有了新的内涵，成为了独特的新古典主义。

2. 新古典风格的特点
2.1 建筑特点

新古典主义的基本元素来源于古典建筑，因此在建筑设计上有很明显的体现。对称和平衡是新古典主义最大的特点，也就是古希腊和古罗马建筑所强调的比例的适当。而新古典主义建筑最明显的组成元素就是柱式的使用。在外部，柱式可以有效地平衡建筑的比例，同时也是古希腊和古罗马建筑的象征性符号，柱式的使用合理不仅可以

Chapter One

OVERVIEW FOR NEO-CLASSICAL HOTELS

Figure 3. The interior was designed by Soufflot to achieve clarity, with numerous windows and slender columns.
Figure 4. In the dome of the Panthéon is a beautiful fresco representing the Glorification of Sainte Geneviève.

图3 苏夫洛设计的先贤祠内部清晰明亮，有许多窗户和细长的柱子。
图4 先贤祠拱顶是颂扬圣·热内维耶瓦的精美壁画。

Pantheon is the master work by French architect Jacques Germain Soufflot, which is originally the church built by the command of Louis XV of France. On the plan, the church is laid in the shape of the ancient Greece cross and its façade follows the temple type, of which the length is 110m, the width 84m and the height 83m to the upper dome (Figure 3.4). The façade imitates Pantheon in Rome, and the gable is supported by the portico with 22 Corinthian columns. The main hall with 32 columns sustains an 83-metre-high dome, which is the most arresting part in the whole building. The central dome is constructed with 3 levels. The bottom level with a hole straight up to the middle, and the top level off the ground is about 70m. The dome design also can trace back to the period of ancient Rome. Pantheon in Rome is still intact today, of which main part is a structure of dome, sculptured by the nature without any support and made of pouring concrete from volcanic ash. It is can be called a miracle of ancient architecture.

Pantheon in Paris is identified as the 使建筑看起来更加壮观，也能为建筑提供有力的支撑以及均衡感。其中，法国的先贤祠（Pantheon in Paris）（图2）可以称为早期新古典主义建筑的典范。

先贤祠又被称做万神殿，是法国建筑师雅克·日尔曼·苏夫洛（Jacques Germain Soufflot）的代表作，原本是受当时的法国国王路易十五之命所兴建的教堂。这座教堂整个平面呈古希腊十字形，立面为神殿式，长110米，宽84米，上部的圆顶高达83米。正面仿造罗马万神殿，由22根科林斯柱组成的门廊支撑着山墙。正厅里是由32根圆柱支撑起的83米高的穹顶（图3、图4），这个穹顶也是整个建筑中最引人注目的部分。中央穹顶有三层结构，最下层顶部开圆洞，直达中层，最高层离地面近70米。这种穹顶的设计也可以

representative in the early period of neo-classicism in France, which inherits the rationality, harmony, elegance and other characteristics from classical architecture, but without the strict compliance to geometry proportion in classical architecture. And it is integrated with Gothic lightness and classical stateliness, representing the new form of architecture. The designer Soufflot was also identified as a representative of neo-classicism in French, who advocated the revival of classical architecture in France when rococo style prevailed, and then introduced neo-classical style into France, making a great contribution to the broadcast of neo-classicism.

2.2 Characteristics of Interior Design

The same as architectural design, interior design in the 18th century tended to revive classical design. Overmuch decor of baroque and rococo were criticised. The excavations to Pompeii and Herculaneum steered the direction of design to accord with the form of neo-classical architecture.

The primary characteristic of neo-classical design emphasises the revival of pure art in the period of ancient Rome, and therefore the classic elements also learned from the relics in the period.

The architectural design in ancient Rome inherited the style in the period of ancient Greece, and was renovated in many respects, reaching a very high level in art. Magnificent and spectacular, by virtue of the aesthetic concept in ancient Greece, the architecture was granted very strong technicality at the same time, in particular, in the use of arch technique. Arch, an arc shaped structure, has the effect of load-bearing and decoration, and semicircular arch is known as the "Rome Arch" (Figure 5). In

追溯到古罗马时期。古罗马万神殿至今仍保存完好，主体部分也是一个圆顶结构，并且整个穹顶浑然天成，没有任何支撑，由火山灰制成的混凝土浇筑而成，称得上是古代建筑的奇迹。

先贤祠被认为是法国新古典主义建筑的早期代表，继承了古典建筑理性、和谐、典雅等特点，但却并没有古典建筑对几何比例的严格遵守，融合了哥特风格的轻巧和古典的庄严，代表了建筑的新形式。先贤祠的设计者苏夫洛也被称为法国新古典主义的代表性人物，他在当时洛可可风格盛行的法国倡导古典的复兴，将新古典主义引入了法国，也为新古典主义的传播做出了巨大的贡献。

Figure 5. Rome Arch. To support the tremendous weight of the arches, it was necessary to provide a way of transmitting the force to massive piers and then to the foundation of the arch. The Romans achieved this feat through the use of the Keystone block. The force was directed down onto the top of the keystone. Because of its shape the force was translated to the voussoir blocks of the arch which in turn translated the force through the impost to the piers. During construction, the voussoir's were supported by a temporary wooden frame until the keystone was inserted.

图5 罗马券。为了支撑拱的巨大重量，必须先把重量传递到墩子上。罗马券的技巧在于拱顶石的运用。拱的重量被直接传递到了上面的拱顶石上。梯形的拱顶石又可以把重量传递到旁边的楔形拱石上，反过来再通过拱端托传递到墩子上。在建造过程中，拱石先由临时的木制框架支撑，拱顶石最后放入。

013

Chapter One
OVERVIEW FOR NEO-CLASSICAL HOTELS

ancient Rome, natural concrete was adopted in the arch construction because of its extraordinary solidness. The arch structure can extend interior space. Appearing in the 1st century, cross-shaped arch can disperse the weight and then focus it onto the 4 corners; hence, the structure that is free from load-bearing walls makes the space more spacious, such as Roman Imperial Thermae, one representative among the architectures on the catalog. In addition, the architectures in ancient Rome created the combination of arch and colonnade. It is a line of continuous arches, of which the feet of arch stand on the entablature of order, such as Sant Apllinare Church. Moreover, abundant frescos were explored from the relics of Pompeii. These lively and vivid frescos show the highest level in art (Figure 6). Besides, sculpting technique was mature in ancient Rome. The main achievement was portrait sculpture which emphasising reality and personality and is full of harmonious beauty. Usually, most of these portraits are static with reserved countenance.

In ancient Rome, architectural art and interior design made high achievements and became the replicated models in the field of neo-classicism. And the most important part was still the use of order, which was the foremost characteristic of neo-classical design. In Pantheon, Rome, the most antique and simple Doric order is adopted, of which scape is with groove and chapter is smooth and mellow. The chapiter of Ionic order is decked with eddy, while, highly decorated, Corinthian order, carved with buttercup patterns is like a basket of flowers. Order is used widely in interior design. Continuous orders in a rhythmic arrangement highlight the strong sense of rhythm and order, as well as the symmetrical effect, which is the indispensable element of neo-classical design.

Furthermore, new forms abstracted from classical components were usually re-organised, such as fresco, sculpture, etc. The frescos excavateded from Pompeii greatly enlighten neo-classical design. Either on the ceiling or the wall, fresco was a

2.2 室内设计的特点

与建筑设计相同，18世纪的室内装饰上也出现了复兴古典的趋势。巴洛克和洛可可过度装饰的风格受到批判，庞贝古城和赫库兰尼姆古城的发掘使室内的设计开始倾向于与新古典主义建筑相符合的形式。

新古典室内设计的首要特点是强调回到古罗马时代的纯粹的艺术形式，因此，其经典元素同样来源于古罗马时期的遗迹。

古罗马的建筑沿袭了古希腊时期风格，并且在许

Figure 6. Fresco in Pompeii
图6 庞贝壁画

popular decoration used in interior design, and sculpture was an indispensable furnishing in neo-classical style. Compared with ancient roman period, the themes of frescos and sculptures were a richer catalog, more consideration taken into account on the coordination with interior atmosphere. However, portrait sculpture and frescos with the themes of Bible stories were still the main option (Figure 7).

Another characteristic in neo-classical design is the emphasis on rationality. From 17th century, normal education about architecture began to emerge. In the past, however, the technologies of construction and design relied on the heritance of craftsman from generation to generation. In the meantime, the essence of scientific rationality advocated in the Enlightenment movement was broadcast widely into varied fields. The logical and economic problems began to be taken into consideration during the process of design, and more rational thoughts were encouraged. If baroque and rococo are inclined to perceptual styles, neo-classical design is a rational art form, or namely a form rooted in rationalism. However, this kind of rationality not only applies and complies with geometric proportion and patterns natural world as well. Its principle highlights logicality, advocates simple and clear style and requires the functionality and rationality (Figure 8).

Figure 7. Syon House, Middlesex, England, 1760 – 1769, Designed by Robert Adam
图7 英国米德尔塞克斯西恩住宅（1760 – 1769），由新古典主义大师罗伯特·亚当设计。

多方面有新的创造，达到了很高的艺术成就。建筑物的风格雄伟壮观，借助了希腊时期的美学概念，同时也有很强的技术性，其中最有特色的是拱券技术的应用。拱券是一种外形为圆弧形状的建筑结构，具有承重和装饰的作用，半圆形的拱券被称为"罗马券"（图5）。古罗马时期的拱券技术使用天然混凝土，异常坚固，通过这种拱券结构可以使室内获得广阔的空间，而在公元一世纪出现的十字拱则可以把顶部重量集中到四角，无需承重墙，能获得更宽敞的空间。如古罗马的皇家浴场就是其中的代表。除此之外，古罗马建筑还创造了拱券与柱列的组合，将券脚立在柱式檐部上的连续券。另外，从庞贝古城的遗址中还发现了大量的壁画，这些壁画栩栩如生，表现出了很高的艺术水平（图6）。除此之外，古罗马时期雕刻技术也很成熟。古罗马雕刻的主要成就体现在人像雕刻上，对人物的雕刻强调真实和个性，且富于和谐美，人物形象多处于静态中，表情矜持。

古罗马时期在建筑艺术和室内装饰上取得了很高的成就，成为了新古典主义所效仿的对象。而其中最重要的依然是柱式的使用。柱式是新古典设计最重要的特征，在罗马万神殿中运用了最古老最简单的陶立克柱式，柱身带有凹槽，柱头光滑圆润。爱奥尼克柱式柱头有涡卷装饰，而科林斯柱式柱头则由毛茛叶纹装饰，像一个插满鲜花的花篮，装饰性很强。柱式在室内的使用范围很广泛，连续而有节奏的柱式使用可以使室内空间形成强烈的韵律感和有序感，也可以产生具有古典美感的对称效果，是新古典设计中不可缺少的元素。

Chapter One
OVERVIEW FOR NEO-CLASSICAL HOTELS

Another characteristic of neo-classical interior design is the rigorous and elegantform. The rigorous form and symmetrical elegance are always the aim that neo-classical interior design pursues. No matter the interface finishing, the furniture style or furnishing art, all are based on the detail elements in the period of ancient Rome and ancient Greece, highlighting the form, adopting linear figure and presenting elegant and grandeur atmosphere. In particular, furniture design prefers to use linear and geometric patterns, returning to simple and plain form from complicatedly decorated rococo style. It was reflected perfectly by the work of Robert Adam, the representative of neo-classicism. Robert Adam was a Scottish architect, as well as the interior designer and furniture designer, whose works have rigorous proportion, classical and delicate decor, expressing the characteristic of neo-classicism. In his classical form design, more bright colours were infused, which made interior space more grandeur and elegant through the contrast of colours and lines. The characteristic was inspired by the relics of ancient Rome. The books about Greek history published in 1860s introduced a great quantity of decor elements, such as the great sphinx, ivy leaves, etc. By means of paintings, gilding and other methods, Adam applied these décor elements when he designed furnishing details like the mirrors, pendant lamp, cornice, etc, to achieve the effect of uniformity. At the same time, Adam abandoned traditional panels and coats the walls with light coloured paint instead. Also, he divided different areas with pouring lines in perfect proportion. Luxurious plas-

另外新古典设计也常把古典构件以新的形式抽象出来加以组合。如壁画、雕塑的使用。庞贝遗址发掘出的壁画给了新古典设计很大的启发，在室内装饰中，无论是天花板还是墙壁，壁画都是一种常用的装饰，而雕塑更是新古典风格中不可缺少的陈设品。相对于古罗马时期来说，壁画和雕塑的题材更加广泛，更多的考虑与室内环境的配合，而人像雕塑和圣经故事题材的壁画依然是重要的选择（图7）。

新古典的室内设计中另一个特点是对理性的强调。从17世纪开始，正规的建筑教育开始出现，而在此之前建筑和设计技术都是依靠工匠的代代相传得以传承。与此同时，启蒙运动所崇尚的科学理性的精神广泛地传播，并应用到各个领域，设计过程开始有了更多关于逻辑及经济问题的考虑，也有了更多的理性思考。如果说巴洛克和洛可可风格强调的是感性的装饰，那么新古典设计则是强调理性的艺术形式，或者说根植于理性主义。而这种理性并非古罗马时期的对几何比例的遵守运用和对自然界的模仿，它的原则是注重设

Figure 8. Great Hall in Syon. Adam employed Joseph Rose to carry out the breath-taking decorative stucco work in the House and in this room; the cool pale tones are only broken by the black and white marble flooring which echoes the ceiling pattern.

图8 西恩住宅大厅。其中精美的粉刷由约瑟夫·罗斯（Joseph Rose）完成，黑白两色的大理石地板打破了整体的浅色调，并与天花板上的图案相呼应。

ter ceiling was adorned with relief or gilding, conveying richness of neo-classicism. Adam's works had a rich catalog, including public architectures, churches, residences, etc (Figure 9).

3. Manifestations of Neo-classicism in Different Countries

Although neo-classical design was widespread in Rome, Italy, which was one of the birthplaces of this style, it also influenced France, Britain, and America etc. Neo-classical design also varies in different countries.

3.1 Italy

As the architecture of ancient Rome inspired neo-classicism, Rome naturally became one of its centres. Born in Rome, 1762, Giuseppe Valadier, a famous Italian architect, designer, urban planner and archaeologist, was the representative of Italian neo-classicism for his bold design style and beautiful carved tables in particular. His tables were made from deluxe marbles and gilded so as to display dignity and wealth. Besides, he was an expert in urban renovation and historical relics repair.

计的逻辑性，提倡简洁、清晰的风格，要求设计的功能性和合理性（图8）。

新古典室内设计的另一个特点是其严谨和优雅的形式。形式的严谨，对称的优雅一直是新古典室内装饰中所追求的目标。无论是界面的处理、家具的风格以及陈设艺术等都基于古罗马和古希腊时期的细部元素，注重形式，线条多为直线形，展示高雅大气的氛围。尤其在家具的设计上，更多的采用了直线和几何形式，从装饰复杂的洛可可风格中回归到简单朴素的形式。这种特点在新古典主义的代表性人物罗伯特·亚当（Robert Adam）的作品上有很好的体现。罗伯特·亚当是苏格兰建筑师，同时也是室内设计师和家具设计师，他的室内设计作品比例严谨、装饰古典精致，充分表现了新古典主义的特点。他在古典形式的设计中融入了更加鲜明多彩的颜色，使室内空间通过线条和颜色的对比更加庄严优雅，这种特点也是受到了古罗马遗址的启发。而18世纪

Figure 9. Built in 1767, the entrance lobby of Osterley Park Housewas satisfied Adam's desire of removing modeling, light and shades in the space. In the square, semicircular niches at both sides are striking, of which either side has a fireplace in which statues are laid. As a multi-functional room, the lobby serves as a living room, reception office and temporary dining hall.

图9 奥斯特雷公园住宅的入口大厅建于1767年。它满足了亚当想让空间中造型，灯光和阴影运动起来的渴望。 矩形的空间中，两端半圆形的壁龛引人注目，它们两侧各有一个壁炉，其中放置着雕像。作为多用途的房间，大厅可作为会客室、接待室和临时餐厅。

Chapter One
OVERVIEW FOR NEO-CLASSICAL HOTELS

The origin of neo-classical furniture is Turin, another Italian city adjacent to France, whose furniture style is widely influenced by its neighbour. In 1770s, furniture manufactured in the region of Piedmont and Turin in particular was assessed as the model of Italian furniture for its elegant design and high-end materials. What's more, many neo-classical buildings scatter in Turin. Taking the French classic, Versailles, for reference, all of them represent Italian neo-classicism.

In Milan, neoclassical design features simplicity and solemnity, and furniture, without exaggerated decorations, is mainly made from walnut. The most famous one is the cupboard designed by Giuseppe Maggiolini, a cupboard manufacturer of the late 18th century whose style centred on late baroque and neo-classicism represented by office desks, dining tables and cupboards. Most of them were produced by imported woods or European's with their original colors and the patterns on their surfaces were all created by well-known artists (Figure 10). Apart from cupboards and tables or desks, chairs were also outstanding. Italian chairs of neo-classical period were similar to those of baroque: bold design, heavy leg and complex carving and vertical lines. It was popular in Venice to have chairs gilded but in Milan nevertheless.

3.2 France
Unlike many other countries, neo-classicism in France is divided into three stages: the Luis XVI style, the Directoire style and the empire style.

Figure 10. A pair of neo-classical marquetry commodes in the manner of giuseppe maggiolini
图10 朱佩塞玛格奥里尼式的新古典镶花洗脸台。

60年代出版的有关希腊时期历史的书籍介绍了大量当时的装饰元素，狮身人面像、藤叶等，亚当通过绘画、镀金等方式把它们用在了自己的设计细节中，如镜子、吊灯、檐口等处，以达到主题统一的装饰效果。同时，亚当在墙面的处理上放弃了传统的镶板而涂以浅色漆，并且用浇注的线条在墙面上按完美的比例分割出不同区域。华丽的石膏天花板上饰以浮雕或镀金，相对于古典设计更有丰富性。亚当的作品有很多，包括公共建筑、教堂、住宅等（图9）。

3.新古典主义风格在各个国家的表现形式
虽然新古典主义设计主要出现在意大利的罗马，但它只是这种风格的发源地之一，18世纪的法国、英国、西班牙、美国等地同样受到了新古典风格影响。并且在不同的国家，新古典风格有不同的表现形式。

3.1意大利
古罗马建筑给予了新古典主义灵感，罗马也就顺理成章地成了这种风格的中心之一。基赛匹·维拉迪尔（Giuseppe Valadier）是其中重要的代表。1762年出生于罗马的维拉迪尔是意大利著名建筑师、设计师、城市规划师和考古学家，他也是意大利新古典风格的主要代表人物，以设计大胆、雕刻精美的桌子著称。他的桌子常用豪华的大理石制成，并且镀金以表现尊贵和财富。同时，他还擅长城市改造和修复古代遗迹。

意大利的另一个城市都灵则是新古典风格家具的产地。都灵临近法国，家具的设计风格也深

French interior design experienced the transformation from late rococo to neo-classicism during the tenure of the King of France, Luis XVI. In reality, the French neo-classical style emerged as early as the reign of Luis XV. But it steered away from the convention and turned to simplicity. Between 17 century and 18 century, the palace style played a leading role in French interior design, in company with the upper class constantly chasing novel lifestyle. Up to late 18 century, the artificiality of rococo could not satisfy the public's curiosity and the architectural style of ancient Rome and Greece were revived, and to a large extent the support from Luis XVI and his wife, Marie Antoinette, paved the way for the Luis XVI style.

Jacques-Ange Gabriel was a French neo-classical architect. His masterpiece was Petit Trianon. Constructed between 1762 and 1768, Petit Triano was the typical example of the shift from rococo to early neo-classicism. Although it covered narrow acres, the entire palace was proportionally harmonious. Its main rooms, decorated with the colour of light browse or white, were laid in a rectangular form with panels and frames shaped in classical arches on the wall (Figure 11). This style had been widely imitated since it was born, pioneering the French design style. At the same time, the delicate textiles by Philippe de Lasalle and the rigorous adornments by Richard de Lalonde also exerted great impact on the Luis XVI style.

Figure 11. Interior of Petit Trianon in Paris
图11 小特里阿农宫室内

受法国影响。18世纪70年代，意大利皮埃蒙特（Piedmont）地区，尤其是都灵生产的家具以其优雅的设计和高端的材料成为意大利家具的典范。另外，都灵也有许多新古典主义风格的建筑，这些建筑借鉴了法国的经典建筑凡尔赛宫，是意大利新古典主义建筑的代表。

米兰的新古典设计以简洁庄重为特点，家具多由胡桃木制成并且没有过多修饰。其中，由Giuseppe Maggiolini设计的柜类最为著名。Giuseppe Maggiolini是18世纪晚期米兰橱柜制造商，他的橱柜风格以晚期巴洛克和新古典主义风格为主，代表性的家具有五斗柜、办公桌、餐桌等，橱柜，大多由进口木材或欧洲木材制成，使用木材的自然颜色，家具上的图案由著名的艺术家设计（图10）。除了柜类和桌类家具外，椅类家具也很突出。新古典时期意大利的椅类家具类似于巴洛克风格，设计大胆，椅腿笨重并带有复杂的雕刻，常用直线条。在威尼斯地区的设计中常用镀金，而米兰则与之相反。

3.2 法国

不同于其他国家，法国的新古典设计分为若干个阶段：路易十六风格、督政府风格（The Directoire style）和帝国风格。

法国国王路易十六在位的时期是法国室内设计经历洛可可风格后期及向新古典风格转变的时期。其实，早在路易十五在位时期，法国的新古典风格已经出现，但它并没有延续路易十五风格的曲线装饰，而是走了更为简洁的道路。17、18世纪的法国室内设计以宫廷风格为主导，上流社会不

Chapter One
OVERVIEW FOR NEO-CLASSICAL HOTELS

Delicate hand-painted wallpaper was another feature of the Luis XVI. Wallpapers can also be designed in marble or wooden textures. What's more, some walls were covered with wallpapers of stripes and geometry, and then rolled edges. In general, bigger rooms hang tapestries.

As the outburst of French Revolution, science and art were greatly enhanced. As a result, interior design was more and more influenced by historical relics. By the end of 18th century, a more simple, more intended to the art of ancient Rome and Greece style, the Directoire style had developed. Georges Jacob, a cupboard designer in Paris, had an impact on this style. Compared with the Luis XVI style, Jacob's was simpler, muted colors usually decorating the walls and tassels, laces in company with textiles (Figure 12). Besides, as the successor of the Luis XV style and the forerunner of the empire style, the Directoire style was influenced by contemporary financial conditions, which can be found in the less utility of rare woods, the

Figure 12. Ceremonial Bedroom of Pauline Borghese by Georges Jacob
图12 乔治·雅各布设计的波林·鲍格才式卧房。

断追求着新颖的生活方式，到了18世纪晚期，洛可可式的矫揉造作已经不能满足人们的好奇心，而古罗马和古希腊的建筑风格重新唤醒了热情，同时，国王路易十六和他的妻子玛丽·安托瓦内特（Marie Antoinette）的支持也是在很大程度上为路易十六风格的发展起了很大作用。

雅克·昂热-加布里埃尔（Jacques-Ange Gabriel）是法国新古典风格的建筑师，他最重要的作品是小特里阿农宫（Petit Trianon）。小特里阿农宫修建于1762-1768年，是洛可可风格向早期新古典风格转变的典型之作。虽然面积不大，但整个宫殿比例和谐，主要的房间都呈长方形，以浅灰色或白色装饰，墙上的镶板和镜框制成古典的拱形（图11）。这种风格出现后随即被大量的模仿，引领了法国的设计风格。同时，室内由法国艺术家菲利普·拉萨勒（Philippe de Lasalle）设计的精致的织物和理查德·兰隆德（Richard de Lalonde）设计的严谨的装饰也对法国的路易十六风格产生了很大影响。

在路易十六风格中，精致的中国手绘壁纸也是一大特色。壁纸也可以设计成大理石纹、木纹等样式，产生建筑设计的效果。另外还有一些把墙面覆盖条纹或几何图形墙纸，然后装饰滚边。较大的房间一般会悬挂大型壁毯等。

随着法国大革命的爆发，科学和艺术的发展受到鼓舞，室内的设计越来越多地受到古代遗址的影响。18世纪末，一种更加朴素、更加倾向于古罗马和古希腊艺术的督政府风格发展起来。巴黎

simpler ornaments, and the omitted intricate wood-mosaic in particular. Because of its imitation of classical design, the Directoire style was more rigid, wider in size, thicker and simpler in appearance (Figure 13).

In 1804, Napoleon proclaimed himself emperor, and so was influenced the design style. One striking feature was that military elements started permeating into interior design. Besides, luxury and rigorous style fully presented regal demeanor. The colours of black and gold were principal hues. In addition, the red walls excavated from Pompeii was very popular. Patterns of symbolised animals were usually used in decorative details and the initial letter "N" can be found everywhere. The most typical architecture of this style is the palace named Palais de Fontainebleau located in Fontainebleau, a town in northern France. It was Luis VI who ordered to construct the palace and his successors like Luis XV, Luis XVI, and Napoleon etc.. all once lived in it. With the passage of time, Palais de Fontainebleau underwent renovations, of which the one conducted in 1808 by French architects Percier and Fontaine employed the classic elements of Napoleon period such as gilded decorations, lofty mirrors, red or green silks, etc.. The symbolised design was vigorous in the line drawing and magnificence in the general layout, fully displays Napoleon's position, power and his

的橱柜设计师乔治·雅各布（Georges Jacob）对这种风格有很大影响，雅各布的设计比起路易十六风格更为简单，常用柔和的色调装饰墙壁，织物带有古希腊细部的花边流苏等（图12）。同时，作为对路易十五风格的承接和对帝国风格的开启，督政府风格受当时财政状况的影响，家具设计减少了对珍贵木材的使用，装饰也相应简化，尤其是复杂的镶嵌细木工。对古典设计的模仿使这种风格比较古板，尺度也比较大，看上去厚重朴实（图13）。

Figure 13.The seven drawer chest of walnut having paneled construction to the case with recessed panel drawers. Fluted columns flank left and right with ribbed brass cuffs accenting to either side of the top drawer. The whole raised on trumpet shape feet with brass ring accents.

图13 这个七个抽屉的胡桃木柜使用了镶嵌工艺，带有嵌壁式的抽屉板，左右两边是带有凹槽的圆柱，最上面抽屉两边的圆柱上还有黄铜装饰。柜子带有黄铜拉手和喇叭形的柜腿。

Chapter One
OVERVIEW FOR NEO-CLASSICAL HOTELS

heroism complex (Figure 14 and 15).

3.3 England

Generally speaking, neo-classicism in England started from the second half of the 18th century and lasted till the early 19th century. England was under the reign of King George in the 18th century, but since the "Glorious Revolution" in 17th century, there had been a long-lasting political struggle and an ever-increasingly fierce conflict between the Parliament and the court. Until 18th century, the Parliament seized the power and replaced the King, the political puppet. Therefore, artistic forms did thrive among the people instead of merely in the court though they were supported by the royal family during the neo-classical period.

With the overthrow of English feudalism and the foundation of bourgeois regime, the English bourgeois accumulated tremendous wealth and labor force through enclosure movement and overseas exploitation. In the meanwhile, England possessed

1804年，拿破仑称帝，艺术风格也因此受到影响。最为明显的是在室内设计中引入了军事题材。整个风格豪华，严谨，充分体现了帝王的风范。色彩上黑色和金色成为主色调，另外，庞贝出土的红墙使这种"庞贝红"也很受欢迎。在装饰的细节中，常用有象征意义的动物图案，拿破仑名字的首字母"N"也随处可见。最能体现这种风格特点的可以算是法国的枫丹白露宫。枫丹白露宫坐落于法国北部枫丹白露镇，由法国国王路易六世下令修建，后来，路易十五、路易十六、拿破仑等都曾在此居住。枫丹白露宫经过多次修葺，其中，由法国设计师佩尔西埃（Percier）和方丹（Fontaine）于1808年重新设计的一些套房采用了拿破仑时期的经典元素，镀金的装饰、高大的镜子、绿色和红色的丝绸等，这些

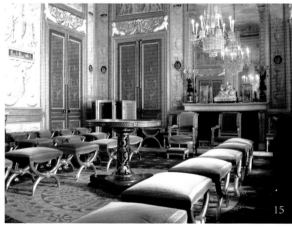

Figure 14. Grand salon of Fontainebleau
Figure 15. The throne room, Fontainebleau

图14 枫丹白露宫的大沙龙
图15 枫丹白露的正殿

workplaces of first-class: large scale, high efficiency, abundant manufacturing techniques and skilled workers, all of which provided indispensable conditions for machine manufacturing. In 1733, John Kay invented "shuttle flying" which greatly promoted the speed of cloth weaving. In 1769, Scottish mechanic James Watt produced the first steam engine. The invention and application of steam engine was the first leap in the history of techniques and was the symbol of the first industrial revolution. Since then, England entered the "Era of Steam", during which the steam engine transfused great drive into industrial revolution. The significant innovation in different production departments profoundly influenced the development of interior design.

With the development of science and technology, the social reforms and the influences by Italy, France and many other countries, England also witnessed neoclassicism movement in the late 18th century. Similarly, neo-classicism in England initially opposed fussily decorated styles of baroque and rococo. Nevertheless, apart from assimilation of ancient Roman and ancient Greek architectural styles, English neo-classical design enjoyed exceptional advantages — novel materials and new technologies.

The art works delegated to represent English neo-classicism were created by Robert Adam. His masterpieces were influential in Britain, Scotland and Russia contemporarily. Since 1754, Robert Adam started a long-time journey, spending five years visiting places of great interest in France and Italy and acquiring the knowledge of architecture. On his return, Robert Adam opened a work studio with his brother, James Adam, engaging in architectural and interior design. At that time, classical buildings were becoming popular, but the brother Adams did not follow the trend and they developed their own, instead. Their art works were influenced by ancient Roman architectures, but unlike Palladianism's the strict implement to classical architecture with clear, straight, rigorousand pristine lines and flat 符号化的设计线条硬朗、宏伟豪放，充分展现了拿破仑的地位、权势以及英雄主义情结（图14、图15）。

3.3英国

一般来说，英国的新古典时期从18世纪后半叶开始，一直持续到19世纪前期。18世纪的英国是乔治王朝的时代，但从17世纪末的"光荣革命"后，英国的政治斗争不断，议会和宫廷之间的冲突愈演愈烈，直到18世纪，议会的权力逐渐取代了国王，国王变成了统而不治的傀儡。因此，在新古典时期，英国的艺术形式并没有因为得到皇室的支持而变得高高在上，反而在民间有了蓬勃的发展。

随着英国封建专制制度的推翻，资产阶级政权的建立，英国的资产阶级通过圈地运动以及海外掠夺积累了大量的财富和劳动力。同时，英国拥有一流的手工工场，规模大、效率高，有丰富的生产技术知识和熟练的技术工人，为工业发展所必需的机器制造创造了条件。1733年，约翰·凯伊（John Kay）发明了"飞梭"，使织布的速度得到了很大提高，1769年，苏格兰机械师詹姆斯·瓦特（James Watt）研制出了第一部蒸汽机。蒸汽机的发明和应用是人类技术史上的一次飞跃，也是第一次工业革命的主要标志。从此以后，英国进入了蒸汽时代，蒸汽机为工业革命注入了强大的活力，各个生产部门都有了重大的革新，对室内设计的发展也产生了很大影响。

随着科技的发展、社会的变革以及意大利、法国

Chapter One
OVERVIEW FOR NEO-CLASSICAL HOTELS

surface. The themes gave priority to elegant and classical and elegant art. The common themes included oval-shaped rose pattern, vertical palm leaves, saucer-shaped decorations, etc.

Besides, carvings on the surface of furniture were exquisite and decoration methods such as painting, gilding, inlay, etc. are employed. And on the subject of materials, mahogany and basswood were preferable.

Because of its elegance and better layout with classical features, the brother Adams' design was popular in London at that time. In addition, Robert Adam suggested new design philosophy, surpassing traditional interior design concepts, and formed the Adam style (Figure 16).

3.4 America
From the 17th century, British started to inhabit in the Americas. Using mud, straw and other materials, they

等国家的影响,英国在18世纪后半期也开始了新古典主义运动。与其他国家相同,英国的新古典设计也是开始于反对巴洛克和洛可可式的繁杂装饰。不同的是,除了对古罗马和古希腊建筑的吸取之外,英国的新古典设计还得天独厚地有了许多新材料和新技术的加入。

最能代表英国新古典风格莫过于前文提到的罗伯特·亚当的作品。他的作品在当时的英格兰、苏格兰及俄罗斯都有很大的影响力。1754年,罗伯特·亚当开始了一次漫长的旅行,他用了5年的时间在法国和意大利参观名胜古迹,并学习建筑知识。回国后,他和弟弟詹姆斯·亚当(James Adam)在伦敦合开了一家工作室,开始从事建筑和室内设计工作。当时古典建筑已经开始流行,但亚当兄弟并没有随波逐流,而是发展出了属于自己的风格,他们的作品受到了古罗马建筑的影响,但并没有像帕拉第奥主义那样对古典建筑的规则严格执行,整体上线条清晰笔直、严谨古朴、表面平坦。装饰题材以古典优雅的艺术主题为主,椭圆玫瑰花饰、垂直棕榈树叶饰、碟形装饰等都是常用的题材。另外,家具的雕刻效果也很精美,并常用漆绘、镀金、镶嵌等装饰手法,材料上常用桃花心木和椴木。亚当兄弟的设计形式优美,有古典特色的同时结构上更加合理,因此风靡伦敦。同时,罗伯特·亚当还提出了超越建筑和室内设计的范围,将室内一切固定的和可移动的物体都包含近设计中的理念,形成了自己独特的亚当风格(图16)。

Figure 16. This Robert Adam's (1728 – 1792) furniture design is an elegant side table with a breakfront marble top which sits above a moulded frieze decorated with ribbon tied swags, centred by a rectangular tablet carved with a classical urn. The turned and tapered, fluted legs are headed by carved flowerhead patarae and stand on turned carved feet.

图16 罗伯特亚当(1726-1792)家具设计系列中的一张优雅的边桌,断层式的大理石桌面安置在装饰着缎带和帷幕的檐带上,中央是一个长方形的雕花。

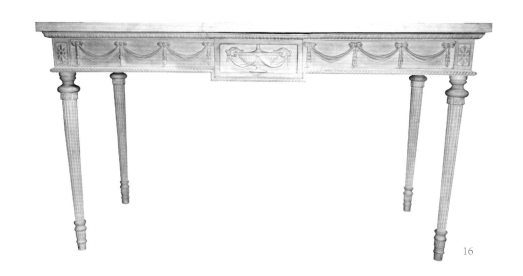

built thatched cottages in the early period. Afterwards, British cottages emerged as the settlement was mushrooming. They were traditional British dwelling houses, solid and practical. Because there were abundant timber resources in the Americas, either floors or ceilings were made from thick boards. This simple and humble style was very practical. Since 18th century, England was under the reign of the Queen Anne and the King George. Consequently, the design style became exquisite and luxurious, and so did the America's. The interior decoration there became richer with exquisite carvings on the panels of the wall and plaster ceilings, added oriental carpets and wallpapers. Later, American handicraft developed to an unparalleled level, and many famous furniture designers popped up. On the basis of the Anne and the George styles and integrated with the characteristics of the America's, the furniture they designed was more agreeable to the American environment, which was great welcomed by many people. In spite of this, these design styles were deeply rooted in the Europe and classified as the early colonial style. In 1776, the Declaration of Independence signed by the thirteen colonies on the American continent proclaimed the colonial period was gone. Correspondingly, the so-called colonial style was out of date. Because the federal government was immediately formed, the American design style was defined as the Federal style, that is, the American neo-classicism, lasting from the late 18th century up to the early 19th century. Likewise, this style lay its foundation on ancient architectures and advocated the revival of ancient Greek and Roman decorative art. Although it was closely related to the European decorative style, the American style was comparatively simple, fresh and practical. The exaggerated decorations such as inlay, carving, gilding, etc. removed, the drawing lines were smoother and the shapes were more unadorned.

As a neo-classical architect, Thomas Jefferson, the third president of the U.S., was probably the most famous neo-classical designer. One of his masterpieces is the House of

3.4 美国

从17世纪开始，美洲大陆有了英国人的定居点，他们用泥、稻草等建造了第一批茅舍。后来，随着定居点的增多，开始出现了英式的木屋，这些木屋是传统的英式住宅，结构坚固，功能性很强。由于美洲大陆木材资源丰富，室内无论是地板还是顶棚，都用厚重的木板制成。这种风格简单朴实，实用性很强。从18世纪开始，英国进入安妮女王和乔治王朝时期，设计风格更加精致豪华，美洲大陆也追随着这种风格，住宅的内部装饰变得更加富丽，墙面的镶板和灰泥的顶棚有了精美的雕刻，还出现了来自东方的地毯和壁纸。到了后期，美国的手工艺变得更加高超，出现了许多著名的家具设计师，他们在英国安妮风格和乔治风格的基础上，结合美洲的特点，制造出来更适合美洲环境的家具，受到很多人的欢迎。但尽管如此，这些设计风格依然根植于欧洲，被划分为美洲早期的殖民地风格。

1776年，美洲大陆13个殖民地签署了《独立宣言》，宣告了殖民地时期已经过去，相应地，殖民地风格一词也不在适用，由于美国刚刚成立联邦政府，因此18世纪末到19世纪初这段时间美国的设计风格被称为联邦风格，也就是美国的新古典风格。美国的新古典风格同样也基于古典建筑形式，提倡复兴古希腊和古罗马的装饰艺术，并且与欧洲的装饰风格有着千丝万缕的联系。但是，美国的新古典风格较欧洲各国的新古典风格更为简洁、清新，也更注重实用性。去除了欧洲新古典风格中过多的镶嵌、雕刻、镀金等装饰，线条更流畅，造型更简朴。

Chapter One

OVERVIEW FOR NEO-CLASSICAL HOTELS

Parliament, Virginia. Six Ionic orders engraved pediments are laid at the House front, which displays the solemnity and elegance. Either side of the House has two layers of windows. Inside there is a hallway, where a statue of Washington is erected (Figure 17). Following the classical architecture style but differing in side windows, the House of Parliament is the representative of the transition to neo-classical architecture style.

Besides, Monticello, Thomas Jefferson's own residence, is also a model of the American neo-classical design. In combination with classical architecture features, the front of Monticello uses Ionic orders and pediments. Inside it, there is a big dome and a winding corridor is at its entrance in connection with rooms on upstairs. Its interior design is delicate and exquisite: the pediments on the door perfectly match with the façade; the carvings on the marble fireplace are exquisite; and the wallpapers are in delicate shades of colors. On the whole, the style of Monticello is elegant and graceful, which fully presents Thomas Jefferson's echoing the best traditions of the classical design and innovating in a new perspective (Figure 18).

在美国的新古典建筑代表设计师中，最为著名的要属托马斯·杰斐逊（Thomas Jefferson）。托马斯·杰斐逊是美国第三位总统，同时也是一位新古典主义流派的建筑师。他最为人熟知的作品是弗吉尼亚州议会大厦。这座大厦的正面用了六根爱奥尼柱式，上面有山花，呈现出了古罗马神庙的庄重典雅。侧面是两层窗户，内部是带有长廊的大厅，大厅摆放着华盛顿的雕像（图17）。整个设计采用了古典建筑的样式，但旁边的窗户又与古典建筑有所不同，因此被看作是向新古典建筑过渡的一座代表性建筑。

Figure 17. The façade of Parliament Virginia
Figure 18. Monticello living room looking toward east entry hall

图17 弗吉尼亚州议会大厦正面
图18 蒙蒂塞洛山庄客厅对着东边的入口大厅

4. Characteristics of Neo-classical Hotels

The neo-classical style is decorated with luxurious decorations, delicate shape and elegant tone, which is suitable for upscale hotels, especially the hotels with large area and space. However, the design of neo-classical hotels requires rich western cultural heritage, full understanding of classical architectures and good command of hotel functions. Hence, compared with the style of modern hotels, neo-classical design is more distinctive.

4.1 Expression of Cultural Connotation

In addition to luxurious appearance, neo-classical design prefers to worship classical culture, appreciate open and free attitude and encourage innovative spirit. In the progress of hotel design, excessively decorative effects often arise, and overemphasis on luxurious atmosphere gives rise to the lack of kernel and connotation in the whole design. Hotel design should not only pursue one-sided luxury, but establishes culture. Excellent hotels should not only make guests feel at home, but lead them to experience unique hotel and local culture. In another word, apart from the good command of the design style, it requires the designer to understand the contents that hotel desires to convey, and to express them accurately. The expression can be visualised, or euphemistic, but must be artistic and acceptable to the guests.

The cultural connotation of hotel requires expressing by integral design. However, hotel has complicated functions. Apart from accommodation environment, dining, leisure and public circulation environment are necessary as well. Hence, the decor for each part needs to coordinate with the overall environment, and then demonstrate the hotel culture better. In the progress of hotel design, the parts such as lobby, restaurants, etc., which are considered as important points, are usually decorated excessively. And corridors, restrooms and so on are neglected, thus causing the skip on design and mismatched environment. It not only affects visual aesthetic, but also im

除此之外，托马斯·杰斐逊设计的自己的住宅——蒙蒂塞洛山庄（Monticello）也是美国新古典设计的典范。蒙蒂塞洛山庄的外观也融合了古典建筑的特点，采用了柱式和山花的形式，内部有高大的穹顶，入口大厅上有回廊，连接上层的各个房间。室内的设计淡雅精巧，门上的山花与外观相得益彰，大理石壁炉雕刻精美，墙面多用浅色的壁纸。整个风格典雅大方，充分体现出了托马斯·杰斐逊对古典设计的传承和发展（图18）。

4.新古典风格酒店的设计特点

新古典风格装饰华丽、造型精美、格调高雅，非常适合高档酒店，尤其是面积和空间都较大的酒店使用。但是，新古典风格的设计需要有浓厚的西方文化底蕴、对古典建筑的深刻理解以及对酒店功能的把握，因此，比起现代风格的酒店设计来说，新古典设计有更多特点。

4.1文化内涵的表达

新古典设计除了华丽的外表外，更注重对古典文化的尊崇、开放自由态度的欣赏以及对创新精神的鼓励，而在酒店设计过程中，往往会产生过分追求装饰效果，过分强调豪华氛围，从而导致整个设计缺少内核，没有思想内涵的现象。酒店的设计不仅仅是对奢华的片面追求，更是文化创造的过程。好的酒店不但能让顾客宾至如归，更能让顾客体验到酒店和当地独特的文化，这就需要设计师除了有对设计风格的把握外，还需要有对酒店所要传达的内容的理解，并准确地表达出来。这种表达可以是直观的，可以是委婉的，但

Chapter One

OVERVIEW FOR NEO-CLASSICAL HOTELS

pacts the continuity of expression. Ballyfin hotel in Ireland, a neo-classical style hotel, after a 9-year renovation, is open for business in 2011. Massive renovation makes every detail delicate in the hotels, and a large number of paintings created by Irish artists in interior space become the most distinctive characteristic, showcasing the art trajectory of Irish artists from mid-17th century to the modern times, as well as reflecting Irish culture, history and social life. Hanging in the lift lobby is the Cootes' portrait, the owner of architecture; showcasing in the bedroom is an Irish painting created in 18-19th century; Adorned in the downstairs bar and treatment room are the works by post-modern Irish artists in 20th century (Figure 19,20 and 21).

4.2 Coordination of Function and Form

The contradiction between function and form is the eternal topic in design, especially in neo-classical style design.

Hotels provide various functions. In the process of design, besides the style for showcasing the aesthetic meaning, the function requirements including the collocation of functions, lighting, illumination, furniture, ventilation and so on are more important. The mould and technology of neo-classical style is comparatively complicated with strong artistry and appreciation. If overmuch is focused on the form, the functionality will be compromised in some

却一定要是艺术的，能被顾客所接受的。
酒店的文化内涵需要通过整体的设计来表达，但酒店的功能复杂，除了住宿环境外，还要有就餐环境、休闲娱乐环境、公共交通环境等，因此，各个部分的装饰需要与整体环境协调，才能更好地表达酒店的文化。而在酒店设计过程中，大堂、餐厅等部分往往是设计的重点而被过度地装饰，走廊、卫生间等处却容易被忽视，从而产生设计上的跳跃，和整体氛围的不搭配，这不仅影响视觉上的美感，更会影响酒店所要表达意义的连续性。

Figure 19. The Stair Hall is brought into a room of its own with its cantilevered stair and hung with Coote family portraits.

图19 回旋楼梯将客人从楼梯间引入一间装饰着库特家族成员肖像的房间

爱尔兰的巴丽芬酒店是一家新古典主义风格酒店，这家酒店经过9年的翻修于2011年正式开业，庞大的翻新工程使这家酒店每个细节都很精致，而其中最大的特点是室内悬挂了众多爱尔兰艺术家创造的油画，这些油画展示了爱尔兰艺术家从17世纪中期开始一直到现代的艺术轨迹，并且也反映出了爱尔兰的文化、历史和社会生活。楼梯间中悬挂的是这栋建筑原来的所有者库特家族成员的肖像，卧室中展示的是18–19世纪的爱尔兰油画，楼下的酒吧和医疗室则布满了20世纪后现代爱尔兰艺术家的作品（图19、图20、图21）。

4.2 功能与形式的协调

功能和形式的矛盾是设计中永恒的话题。在新古典风格设计中，这种矛盾更加突出。酒店本身具有多种功能，在设计过程中，除了展现具有审美意义的风格外，酒店功能配置的合理、良好的采光、合适的照明、舒适的家具、良好的通风等功能性要求更加重要。新古典风格的造型、工艺等相对繁琐，艺术性和欣赏性较强，如果过于强调就会在一定程度上损害功能上的要求，例如，新古典风格所需要的精美厚重的窗幔、床上的帷幔容易阻挡光线、影响通风、造成灰尘的沉积等问题。而一些区域功能上的要求又会在某些方面削弱装饰的效果，造成风格的不连贯。例如客房内卫生间的设计。客房中卫生间是设计的重点，对舒适度、便捷性以及卫生要求等都较高，因此，即使是在新古典风格的设计中，对这部分的设计也常常会化繁为简，更侧重于对功能的要求而直接采用现代的设计。在巴丽芬酒店的客房卫生间

Figure 20. With its gracefully rococo stucco work ceiling, the Lady Caroline Coote Room (formerly Lady Coote's boudoir) is among the most pleasingly elegant bedrooms at Ballyfin. The decoration with a vivid, electric blue wallpaper transforms the room into a suggestive tent-like enclosure. It perfectly reflects the interest in textiles seen in interiors of the Empire period. The colour serves as an ideal foil for the early-Georgian portrait over the chimneypiece showing Henry Meredyth of Newtown, County Meath, by the Irish artist Charles Jervas.

Figure 21. The relaxed atmosphere is perfect for a quiet drink and the chance to admire some of Ballyfin's collection of contemporary Irish art which includes a major work by Michael Farrell (1940-2000) showing Vinent Van Gogh painting at Arles and a vividly expressionist piece above the bar by Brian Maguire which contrast with the gentle lyricism of a series of flowerpieces by the young Irish artist Michael Canning.

图20 优雅的洛可可式灰泥屋顶让卡洛琳·库特女士客房（前库特女士闺房）成为巴丽芬酒店最优雅的客房。客房内装饰着活泼的铁蓝色壁纸，让房间变成了一间犹如帐篷似的空间。这完美地反映帝国时期室内装饰中重视软装的特点。色彩在这里成为壁炉架上的早期乔治王时期肖像画的完美烘托，这幅作品是由爱尔兰艺术家查尔斯·杰瓦斯创作的米斯郡纽镇的亨利·梅莱迪斯的肖像画。

图21 轻松的氛围最适合安静的品酒，借此机会还能充分的欣赏巴丽芬酒店内当代爱尔兰艺术家的精选作品，主要作品包括由迈克尔法瑞尔创作的一幅展现梵高在阿尔斯的作品，以及一幅生动的由布莱恩马奎尔创作的表现主义作品，这与年轻的爱尔兰艺术家迈克尔甘宁创作的文雅的抒情主义花卉画形成鲜明的对比。

Chapter One
OVERVIEW FOR NEO-CLASSICAL HOTELS

extent. For example, neo-classical style requires delicate and massive curtains which can block light, influence ventilation, deposit dust and so on. But, the requirements on function will weaken the effect of decor and impact the consistency of overall style. Take the bathroom design in guestroom for instance. As the centre point in guestroom, the bathroom design has higher requirements on comfort, convenience and hygiene. Hence, even in the neo-classical style design, the part is simplified from complicated design, and emphasis is placed on the function requirements directly by the way of modern design instead. In Ballyfin, the bathrooms in guestrooms solve the problem better. The walls in bathroom adopt the same palette of guestroom, and are adorned with paintings as well. The bathtub and hand washing basin is carved with simple patterns to highlight neo-classical style. Meanwhile, the chandelier above adds the finishing touch and sense of luxury for the space. The overall design is complemented by the style of guestroom, fresh and neat with completed function, which is a successful sample (Figure 22).

中，这个问题得到了比较好的解决。卫生间的墙面采用了和客房相同的色调，并且也同样用了油画做装饰，浴缸和洗手池的表面带有简洁的雕刻，凸显了新古典风格，而上方的水晶吊灯画龙点睛，为空间增添了华贵之感。整个设计与卧室的风格相得益彰，又清新整洁，在功能上也没有缺失，是一个成功的典型（图22）。

Figure 22. The bathroom has every facility that guests could possibly require and the level of luxury throughout is unmatched by any other property in Ireland.

Figure 23. Delightful garden building

图22 浴室具备全套客人需要的设施，奢华等级是其他爱尔兰酒店不能比拟的

图23 优美的花园建筑

4.3 Coordination for Each Part Design

Besides interior design, hotel design involves many other aspects, such as master plan, architectural design, landscape design, etc. Due to each part is in the charge of different designers, it will generate the lack of coordination, and the individud parts in design can be disharmonious with each other. Especially in neo-classical design, as modern hotel architectural design uses modern materials and forms, although the moulds are various and diversified, modern hotel design is incompatible with neo-classical interior environment, as well as exterior landscape design. These deficiencies influence the master plan and image of hotel. Besides neo-classical interior design, Ballyfin itself adopts neo-classical style. The façade uses four Ionic columns, simple and solemn. Two-layer building and walls around are built by stones, natural and classical. In front of the building is a large area of neat lawn. The beautiful environment makes the building like a castle. In the meantime, it can be integrated with neo-classical design as one, and serve as the foil for each other (Figure 23).

4.3各部分设计的统筹

酒店的设计除了室内之外还涉及许多方面，如总体规划、建筑设计、景观设计等。由于各部分设计常常由不同设计单位负责，因此会导致缺少统筹管理，各部分设计不和谐的现象。在新古典设计中尤为如此，现代酒店建筑设计多用现代化材料和形式，虽然造型各异、五光十色，但却与新古典的室内环境格格不入，而室外的景观设计也往往是容易被忽视的地方。这些设计上的缺失影响了酒店的整体规划，也影响了酒店的形象塑造。巴丽芬酒店除了新古典的室内设计之外，酒店建筑本身也采用了新古典风格，正面采用四根爱奥尼克柱式支撑着山花，简洁肃穆，两层的建筑和四周的墙全部用石头砌成，天然古朴，建筑前面是大片的整齐的草坪，优美的环境使建筑宛如城堡一般，同时也能与室内的新古典设计融为一体，互相衬托（图23）。

Chapter Two

SPATIAL LAYOUT OF NEO-CLASSICAL HOTELS
新古典酒店的空间分布

Chapter Two
SPATIAL LAYOUT OF NEO-CLASSICAL HOTELS
新古典酒店的空间分布

Hotel space is generally constitutive of two parts: the front area and the back of house. The front area is an area for receiving the guests where the reception area, the accommodation area, the dining area, the public area, etc. are included. The back of house provides an area for executive work where the administration office, the engineering management area, etc. are included. In the hotel design, the front area design is more important for its various functions.

The reception area in the front area includes entrance, lobby, reception desk, lounge, etc. Apart from adhering to the universal principles of the hotel design, the reception area of the neo-classical hotels possesses distinctive features of their own. For example, lobby is usually spacious with noble, elegant and luxurious style. Therefore, tall classic orders often embellish around the area, which not only play a role of load bearing, but embody a sense of or-

酒店的空间构成大概分为前台和后台两个部分，其中，前台是接待顾客的区域，包括接待区、住宿区、餐饮区、公共活动区等；后台是提供后勤工作的区域，包括办公区、工程管理区等。由于功能的不同，在酒店的设计中，前台各部分的设计更为重要。

前台的接待区包括入口、大堂、服务台、大堂吧等。新古典酒店的接待区设计除了要遵循酒店设计的一般原则之外，还有一些自己的特点。例如，大堂一般空间开阔，强调高贵、典雅、华丽的气派，因此在设计中，四周常用高大的古典柱式做装饰，不仅在结构上能起到承重的作用，也能在宽敞的大堂空间中体现出秩序感，同时，这种新古典风格常用的元素也能很好的分割空间。为了突出新古典风格的尊贵华美，高档的大理石是比较适宜的材料，大面积的大理石搭配大面积的印花地毯是新古典风格的大堂中最常用的组合。但是，新古典设计的元素较多，且颜色较为沉重，因此在大堂中，需要对采光和照明的要求更为严格。

The stately marble columns and gilded gold ceiling and trim of the hotel Lobby and Crystal Foyer welcome guests.

庄严的大理石柱、鎏金天花板、整洁的酒店大堂和水晶门厅欢迎着宾客。

Chapter Two

SPATIAL LAYOUT OF NEO-CLASSICAL HOTELS

der in capacious lobby. Meanwhile, the common element adopted in the neo-classical hotel design can divide space perfectly as well. To highlight the luxury and beauty in the neo-classical style, superior marble is regarded as a more feasible material. Therefore, it is the most common combination in the neo-classical lobby design that a huge area of marble matches with extensive neo-classical printed carpets. But, the design elements of the neo-classical style are rich and their palette typically tends to be dense. Therefore, the requirements of lighting and illumination in lobby design are stricter.

The accommodation area is essential in the hotel. Generally, a great number of neo-classical hotels are luxurious hotels; therefore, the design of luxurious suites is apparently important. Guestrooms provide lodging space, demanding more comfortable and quiet environment, while, as various decorations and modeling are needed in the neo-classical hotel design, it requires choices to be made during the process of the guestroom design. Hence, the choices for furniture, fabric and ornaments are inclined to be delicate and simple ones, and for palettes to harmonious and dynamic ones. Furnishings of the Louis XVI style are comparatively simple and can also showcase luxurious atmosphere; dignified curtains made by superior fabrics are indispensible; canopy bed with draperies that is often adopted in the neo-classical style can be replaced by more simple bed and silk textile; carved elaborately plaster ceiling and chandelier are also the important element for creating the neo-classical atmosphere. In addition, in neo-classical style guestroom, the application of modern appliances such as television, telephone, etc, is a problem. These appliances are necessary in the modern hotel, but incompatible with the neo-classical atmosphere. Thus, the designer needs to hide or weaken the sense of their existence tactfully in the process of design for creating a harmonious environment. In normal condtions, this can be achieved by the selection of the palette, the arrangement of articles and other measures.

住宿区是整个酒店中最基本的区域，一般来说，新古典风格的酒店多为奢华型酒店，所以，高档套房的设计就尤为重要。客房是为旅客提供的住宿场所，要求环境更加舒适安静，而新古典风格的塑造需要较多的装饰和造型，这就要求在客房的设计过程中有所取舍，对家具、织物、装饰品的选择精细而不繁琐，色彩搭配协调而不呆板。路易十六风格的家具相对简单而又能体现出奢华的气息、高档面料制成的厚重窗帘必不可少、而新古典风格中常用的带有帷幔的四柱床则可以用更加简洁的床和丝绸的床品代替、精心雕刻的石膏天棚和水晶灯也是营造新古典氛围的重要元素。此外，在新古典风格的客房中，一些现代设施的应用是设计中的一个难题，如电视、电话等，这些设施在现代酒店中必不可少，但又与新古典的氛围格格不入，因此，需要在设计中巧妙地隐藏或者削弱它们的存在感来达到环境的和谐，一般来说，通过色彩的选择或者位置摆放等手段可以达到这样的效果。

酒店的餐饮区包括餐厅、酒吧、宴会厅等。在面积较大的餐厅或宴会厅中，新古典风格可以很好地呈现，而在面积狭小且比较现代的酒吧中，一般较少采用新古典风格。新古典风格宴会厅中，大型的水晶灯是不可缺少的，造型复杂的水晶灯可以增添宴会厅的华丽氛围，另外，四周墙面可以装饰带有雕刻的镶板或者大型壁毯壁画等，以减少大面积宴会厅的空旷之感。而在面积相对较小的餐厅中，有历史感的油画是最好的装饰品。新古典风格的餐厅多选用圆形桌，因此在摆放上既要合理地利用空间，又要保持中间过道的通畅。

The dining area in the hotel includes restaurants, bars, ballrooms, etc. In relatively large restaurants and ballrooms, the neo-classical style can be presented perfectly, while in the comparatively small and modern bars, the style is rarely adopted. In neo-classical style ballroom, large chandeliers are necessary. Chandeliers with complicated forms can boost the luxurious atmosphere in the ballroom. Furthermore, walls can be set with carved panels or large tapestries and frescos in order to lessen the sense of hollowness. In relatively small restaurants, historical frescos are the best decorations. The neo-classical style restaurants prefer to adopt round tables; therefore, it should utilise the space reasonably in the table arrangement. Simultaneously, good circulation is supposed to be kept in the passages among the tables.

The public area consists of conference rooms, multi-function room, etc. which are normally spacious and appropriate for the neo-classical style, and are expected to meet the requirements of conferences and events. For example, in the conference room, more lighting is needed above the conference table which results in other lighting facilities arranged besides chandeliers. In addition, due to its function, the conference room should not be decorated redundantly in order to avoid participants' distraction. In its decoration, it is advocated to create a neo-classical style environment with concise gimmick. In smaller conference rooms or multi-function rooms, the fireplace possibly serves as the decoration.

公共活动区包括酒店中的会议室、多功能厅等。这些区域一般空间较大，也很适合新古典风格的装饰，但同时也要满足举行各种会议和活动的要求。如在会议室中，会议桌上方需要更多的照明，因此需要安排除了水晶灯之外的照明设施。另外，会议室空间由于其功能要求，界面不宜做过多装饰，以免影响参加会议人员的注意力，在装饰中需要尽量用简洁的手法营造新古典的环境。在相对较小的会议室或多功能厅中，也可以使用壁炉做装饰。

The vast lobby, surrounded by columns and lit by gleaming chandeliers, is the very picture of Edwardian sumptuousness.

宽敞的大堂四周环绕着立柱，采用闪耀的吊灯照明，尽显爱德华时代的奢华氛围。

Reception Area
接待区

The Coexistence of Functionality and Integrity
功能性和整体性共存

There often exists a myth that it cannot achieve a balance between superficial magnificence and inherent functionality. Thus, many designers blindly focus on designing a high, noble and gorgeous lobby, but destroy the functionality and integrity of it. The entrance to the reception area is a place connecting the interior and the exterior; the reception desk undertakes a role for welcoming guests; the spacious area of the lobby usually serves as a lounge for the guests. Besides, the lobby links other areas and steers guests. Hence, these parts in the lobby are not only expected to present design features and be impressed, but to emphasise services. The entrance should be spacious and tidy, and can naturally transit to other areas. Comfortable furniture and art pieces should be adopted; tall green plants are the main elements for adjusting the atmosphere of space. Meanwhile, clear signs are greatly needed to guide guests to other areas, and correspondingly, the lift lobby, the staircase and other areas are expected to be bright and spacious as well.

The reception area is the core of the hotel design, and is the foremost part presenting the hotel design style. Its design style lays the foundation of the whole design keynote; hence, more distinct design features are needed. As the neo-classical style has its own elegant and magnificent qualities, huge areas of reception give it full play.

In the Fairmont Palliser, upon entering, any guest is welcomed by a spacious crystal hall. Around the lobby is lined with stately marble pillars; gilded ceiling is high and noble; floors are paved with blue printed carpets, causing interplay between it and the blue sofa in the lounge. The carpet on the staircase also adopts the colour, which not only demonstrates the design integrity, but functions as guidance for the guests. The lobby links the reception desk, the restaurant and the ballroom with sufficient space and luxurious decorations, adequately showcasing the neo-classical style and satisfying the basic function.

酒店设计中常常有这样的误区，大堂盲目地追求高大、气派、富丽堂皇，而影响了接待区的功能性和整体性。接待区的入口是连接室内室外空间的区域，服务台承担着迎接宾客的作用，而大堂内宽敞的区域一般用于宾客的休息，另外，大堂还起着连接其他各区域以及为顾客指引方向的作用，因此，这几个部分不仅要体现出设计特点，给顾客留下深刻的印用，更要注重服务性。入口处宽敞整洁，过渡自然，大堂休息区采用舒适的家具和艺术品，高大的绿色植物也是调节空间气氛的重要元素。同时，更要有明确的标识引导顾客进入其他区域，相应地，电梯间、楼梯等也应该宽敞明亮。

接待区是酒店设计中的核心，也是体现酒店设计风格最重要的部分，它的设计风格奠定了整个酒店的基调，因此需要更明确的设计特点。新古典风格本身带有优雅富丽的气质，在大面积的接待区域可以有很大的发挥空间。

在帕里瑟尔费尔蒙特酒店中，入口是宽敞的水晶门厅，大堂周围排列着庄严的大理石柱，鎏金天花板高大宏伟，地面铺着大面积的蓝色印花地毯，与休息区的蓝色沙发交相辉映。楼梯的地毯也采用了这种颜色，既体现了设计的整体性，同时也给顾客以导向作用。大堂连接着前台、餐厅和宴会厅，空间充足，装饰奢华，在充分展现了新古典风格同时，也满足了基本的功能。

The Fairmont Palliser

帕里瑟尔费尔蒙特酒店

Location: Calgary, Canada
Designer: W.S. Maxwell
Photographer: Fairmont Hotels & Resorts
Area: 27,000m²

项目地点：加拿大，卡尔加里
设计师：W.S.麦克斯维尔
摄影师：费尔蒙特酒店集团
项目面积：27,000平方米

Centrally located in downtown Calgary, The Fairmont Palliser is conveniently situated near the city's business and financial district and is within walking distance to the city's most exciting retail shopping, arts, culture and entertainment venues.

The spaces have been equipped with oak panels, candelabra, marble columns and floors, handmade carpets and public art. It is a reminder of the golden era of travel – an age of elegance and charm in a classic surrounding, where white-glove service was customary and grand hotels defined a country. The Fairmont Palliser keeps this memory alive through its exceptional service and elegant decor.

The Palliser was built in the Edwardian style commercial. Architect Lawrence Gotch Chicago adopted a look, with geometric lines resembling the countless grain elevators Prairie in Alberta and other prairie provinces. Each of the 405 rooms at the Fairmont Palliser are luxuriously decorated and elegant.

Grand, elegant, and luxurious describe the Heritage Suites at The Fairmont Palliser. These distinctive accommodations are reminiscent of the golden era of travel and are elegantly decorated with ornate architectural details throughout. Heritage Suites feature a pantry area with fridge and microwave, a dining area and ornamental fire place. Guests are welcomed into a spacious foyer area that opens into a separate living area with dedicated guest bathroom. The master bedrooms are substantial in size, luxurious and comfortable, with their own private bathroom. Heritage Suites connect to other Fairmont rooms to form 2 or 3 bedroom suites. Panoramic views of the city are afforded through large windows. The luxurious Royal Suite at The Fairmont Palliser was home to Queen Elizabeth II during her trip to Calgary in June 1990. Located on the Fairmont Gold floor, this distinctive suite has a large bedroom, livingroom, two bathrooms, and a regal dining room which seats 12 guests.

1. The Oak Room is an elegantly casual lounge – a great place for entertaining clients or meeting friends.
2. Original artist mural and landmark heritage fireplace adorn the room with a western flair.

1. 橡木厅设计优雅、轻松，适合宴请宾客或好友聚会。
2. 原创艺术壁画和历史悠久的标志性壁炉为房间增添了西方格调。

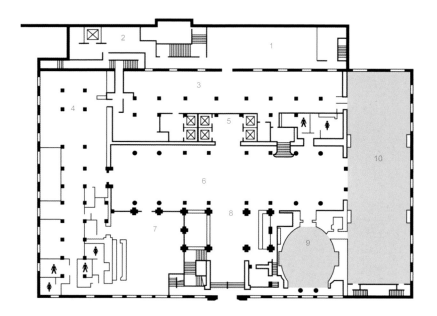

Lobby level plan
1. Kitchen
2. Receiving
3. Food service
4. Rimrock dining room
5. Elevators
6. Lobby
7. Oak room lounge
8. Reception desk
9. Oval room
10. Crystal ballroom

大堂平面图
1. 厨房
2. 上菜处
3. 食品服务
4. 悬崖餐厅
5. 电梯
6. 大堂
7. 橡木酒廊
8. 前台
9. 椭圆厅
10. 水晶宴会厅

1. With 15 crystal chandeliers sparkling from the ceiling, the pale blue and gold accents require no enhancements.
2. The elegantly bar
3. Taking its name from its oval structure and domed ceiling, the Oval Room features a fireplace, crystal wall sconces, and a chandelier.

1. 15盏水晶吊灯在天花板闪烁，点缀的浅蓝色和金色完美无缺。
2. 优雅的酒吧
3. 椭圆厅的名字来源于它的椭圆形结构和拱形天花板，以壁炉、水晶壁灯和吊灯为特色。

1. A historically elegant room with beautiful, understated, hand-painted murals around the walls and ceiling, the Alberta Ballroom is ideal for mid-size to large functions.
2. Architectural detail
3. Gilded gold ceiling and trim of the hotel Lobby and Crystal Foyer welcome guests
4. Grand, elegant, and luxurious describe the Heritage Suites at The Fairmont Palliser. These distinctive accommodations are reminiscent of the golden era of travel and are elegantly decorated with ornate architectural details throughout.

帕里瑟尔费尔蒙特酒店位于卡尔加里市中心，紧邻城市的商业区和金融区，前往市内的购物、艺术、文化和娱乐场所都十分方便。

酒店配有橡木镶板、枝形大烛台、大理石柱和地面、手工地毯以及公共艺术品。它令人想起了旅行的黄金时代——优雅、迷人；古典背景中的一流服务贴心周到，而宏伟的酒店能为国家增色不少。帕里瑟尔费尔蒙特酒店通过接触的服务和典雅的装饰维持了这种鲜活的记忆。

酒店设在一座爱德华七世风格的商业建筑内。建筑师劳伦斯·高奇运用几何线条来模仿了亚伯达大草原和其他草原上不计其数的谷物升降机。酒店的405套客房都采用了奢华优雅的装饰。

酒店的遗产套房以宏大、优雅和奢华为特色。这些独特的住宿空间让人回想起旅行的黄金时代，到处都采用了华丽的细部装饰。遗产套房的备餐区设有冰箱和微波炉，还拥有一个就餐区和装饰性壁炉。客人从宽敞的门厅区域进入带有浴室的独立起居区。主卧异常宽敞、奢华和舒适，配有私人浴室。遗产套房与其他套房连接起来，可以形成2-3卧套房。人们可以从大窗中看到城市美景。

奢华的皇室套房曾在1990年6月为伊利莎白二世女王服务。这间独一无二的套房位于酒店黄金楼层，拥有大卧室、客厅、两间浴室和一间可容纳12名贵宾的皇室餐厅。

1. 亚伯达宴会厅的墙壁和天花板上装饰着美丽低调的手绘壁画，十分适合举行大中型宴会。
2. 建筑细部
3. 镀金天花板和排列整齐的酒店大堂和水晶门厅迎接着宾客。
4. 帕里瑟尔费尔蒙特酒店的遗产套房以宏大、优雅和奢华为特色。这些独特的住宿空间让人回想起旅行的黄金时代，到处都采用了华丽的细部装饰。

Accommodation Area 住宿区

Scenery is Everywhere

风景无处不在

The basic function of a hotel is to meet the accommodation needs; therefore, the design for the accommodation area is another core in the hotel design. But, it is not confined to accomodation under the development of the modern hotel. It also needs to satisfy the psychological requirements of the guests like comfortable sensation, artistic feeling, the embodiment of the status, etc. Thus, personalisation and detail design of guestrooms become more important. Personalisation presents the design concept and idea of the hotel. Different designs for different types of rooms also present the requirements of different guests, and the focus on detail design can add more comfort to guests, as high standard demands for detail design usually make guests feel at home.

If the neo-classical style is designed in an inappropriate way, it will become formalistic. In an era when personalisation is stressed, more changes and connotations that will meet the guests' longing for the neo-classical style are needed, and on the other side, the modern methods that will meet the requirements for comfortable life are also necessay. The coordination between these two sides is important and difficult in the design.

Beau-Rivage Palace is a hotel combining classical and modern ingredients, comfort and personalisation as well. Slightly different as each guestroom, they are all granted with classical elegance, modern comfort and deliberate details. In the guestrooms of Beau-Rivage Palace, scenery is everywhere: gorgeous and noble curtains, paintings from masters, carpets from the Orient, empiral style furniture, exquisite and delicate art pieces on the table, each of which has its own unique history and story. The combination of these details produces senses of art and space, satisfying the function requirements as well as guests' spiritual desires.

酒店最基本的功能是满足住宿需求，因此住宿区的设计也是酒店设计中的核心。然而现代酒店发展已经使它不仅仅满足于一个住的功能需要，还要满足顾客的心理需求，舒适感、艺术感以及身份地位的体现等。因此，客房的个性化和细节的处理变得更加重要。个性化体现了酒店的设计理念和设计思想，不同类型房间的区别设计也体现了不同顾客的需求，而对细节的关注则可以增添顾客的舒适感，宾至如归的感受多半来源于对设计细节的高要求。

新古典风格的设计如果运用不当会产生呆板的感觉，在崇尚个性化的现代，需要有更多的变化和内涵满足顾客对古典的憧憬，也需要现代化的手段满足人们对舒适生活的要求，这两者的协调是设计中的重点和难点。

美岸大酒店是一家集合了古典与现代、舒适与个性的酒店，这里的客房每一间都不尽相同，但却都有着古典的优雅、现代的舒适以及经过深思熟虑的细节。在美岸大酒店的客房里，随处都可以看到风景：华丽厚重的窗帘、出自名家之手的油画、来自东方的地毯、帝国风格的家具、桌上玲珑剔透的艺术品，每一件都有着与众不同的来历和故事。而这些细节组合在一起产生的艺术感和空间感使客房满足了它应有的功能需求，也兼顾了顾客的精神渴望。

Beau-Rivage Palace
美岸大酒店

Location: Lausanne, Switzerland
Designer: Jost, Bezencenet & Schnell
Photographer: Beau-Rivage Palace
Area: 9,200m²

项目地点：瑞士，洛桑
设计师：JBS公司
摄影师：美岸大酒店
项目面积：9,200平方米

The Beau-Rivage Palace is a historical luxury hotel in Lausanne, Switzerland. It was initially built in 1861 and has preserved its original, opulent belle epoque aesthetic throughout the centuries. The palatial architecture combines Italian Renaissance influences with turn-of-the-century baroque touches, which were added when an extension was built to the original palace in 1908.

The interior spaces are a stunning mix of classical and modern design. In response to an ever-increasing demand, construction of the Palace was entrusted to the architects Jost, Bezencenet & Schnell. Their concept of a building that was exuberantly neo-baroque, with later art deco flourishes, is linked to the Beau-Rivage a rotunda of classical simplicity.

There was no question, of course, of saving on quality. The central dome of the main dining room is by Dickmann.

The Zurich painter, Haberer, created the fresco in the Rotonde.

The bronze chandeliers and ornamental wall lights were supplied by Thiébaud of Paris.

The stylist Chiara was responsible for the design of the glass panels which decorate the lift and which are identical to those in the Paris Ritz. Neo-baroque is known for its highly ornamental decorative styling which certainly gave great pleasure to the hotel guests of that period.

18th century frescos, were, from the very start, the scene of sumptuous balls and dinners. The exceptionally beautiful surroundings, the luxurious room furnishings and the highest standards of service soon gave the hotel an international reputation. And such was its almost immediate popularity that it quickly came to the point where the hotel had to expand.

After the construction of a Lobby Lounge overlooking the garden and the lake and the opening of the gourmet restaurant with Anne-Sophie Pic, the "Heart of the Hotel" renovation programme finishes with the construction of a new area for guests' breakfasts.

1. The hotel recently completed an approximately $11.5-million renovation of La Rotonde and created a new connecting glass-enclosed space, La Terrasse.
2. The lobby is large and impressive with beautiful traditional furniture and marble columns which combine to create a regal scene.
3. The Colony Lounge bar boasts an impressive colonial style interior and is a reference.

1. 酒店新近对圆顶餐厅进行了耗资近1,150万美元的翻修,并创建了一个玻璃连接空间——露台厅。
2. 宽敞的酒店大堂配有美观的传统家具和大理石立柱,营造了皇室般的氛围。
3. 殖民地酒廊以独特的殖民地风格装饰为特色,是同类设计中的典范。

Hotel plan
1. Sandoz salon
2. Rotonde salon
3. The terrace
4. Leman salon
5. Hermitage salon
6. Roman salon
7. Beaux-Arts salon
8. Olympic salon
9. Elyes salon
10. Forum salon
11. Atrium
12. Arcades
13. Library (open)
14. Library (closed)
15. Palace terrace
16. Beau-Rivage terrace
17. Beaux-Arts terrace
18. Reception
19. Boutique shop
20. Lobby lounge
21. Spa Cinq Mondes
22. Le bar
23. Miyako
24. Café Beau-Rivage

酒店平面图
1. 山度士沙龙
2. 圆顶沙龙
3. 露台
4. 情人沙龙
5. 隐士沙龙
6. 罗马沙龙
7. 好艺术沙龙
8. 奥林匹克沙龙
9. 伊莱斯沙龙
10. 论坛沙龙
11. 中庭
12. 拱形游廊
13. 图书室（开放）
14. 图书室（封闭）
15. 宫殿露台
16. 美岸露台
17. 好艺术露台
18. 前台
19. 精品店
20. 大堂酒廊
21. 时尚世界水疗中心
22. 酒吧
23. 美弥子餐厅
24. 美岸咖啡厅

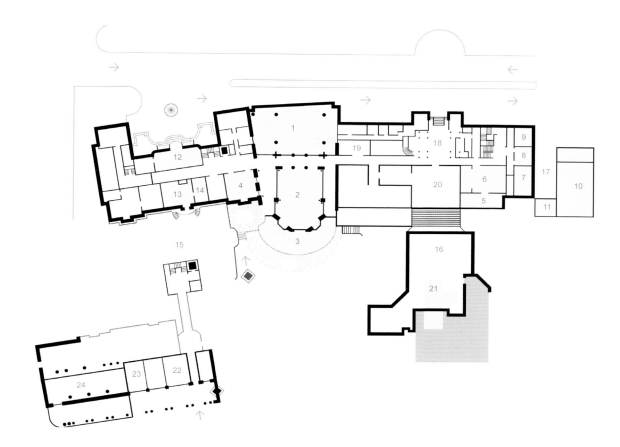

美岸大酒店是瑞士洛桑一家传统的奢华酒店，始建于1861年，保存了它原始而丰富的一战前好时代气息。这座宫殿式建筑采用了意大利文艺复兴风格，并且于1908年扩建时增添了19-20世纪之交的巴洛克风情。

室内空间结合了古典和现代设计。为了应对不断变化的客户需求，酒店委托JBS公司对酒店进行了翻新。他们的建筑主题是生机勃勃的新巴洛克风格，采用了后装饰艺术装饰，与美岸大酒店简洁古典的圆形大厅结合在一起。

毫无疑问，酒店保留了原始的品质。圆顶餐厅由迪克曼设计。苏黎世画家哈勃尔创作了壁画，而铜质吊灯和装饰壁灯则由巴里的蒂挨鲍德供货。

由设计师齐亚拉负责设计装饰电梯的玻璃板与巴黎丽兹酒店的一模一样。新巴洛克主义以其高度的装饰风格而著称，为当时的酒店住客提供了最大的愉悦。

从一开始，华丽的舞厅和餐厅内就装饰了18世纪的壁画。无与伦比的周边环境、奢华的客房装饰和最高等级的服务为酒店赢得了盛名。因此，酒店急需进行扩建。

在建造了一个俯瞰花园和湖泊的大堂休闲吧和安·苏菲美食餐厅之后，酒店的核心翻新工作终于以为宾客提供全新的早餐区域而告一段落。

1. The banquet has baroque style.
2. Le Lobby Lounge and the bar have a relaxed and lounge atmosphere; in the evening, a cosmopolitan decor creates a unique ambience.

1. 宴会厅采用了巴洛克风格。
2. 大堂酒廊氛围轻松休闲；夜晚的国际性装饰营造出独特的氛围。

1. The Junior Suites, facing the lake, feature a sitting-room as well as the latest trends in equipment and technology.
2. This suite of 140 sqm is situated in the central section of the Beau-Rivage wing on the second floor. In Directoire style, it offers a classic image and refined furnishing. Main colour is blue.
3. In shades of mainly red and green, the Royal Suite immerses its guests in a sumptuous world where time seems to have stopped.
4. The bedroom is decorated in a classical way in mainly light shades accentuating the feeling of space and light.

1. 普通套房朝向湖面，配有客厅以及最先进的设备和技术。
2. 套房面积140平方米，位于美岸酒店二楼的中心位置。执政内阁时期的风格赋予了套房古典的形象和精致的家具。套房以蓝色为主色调。
3. 皇室套房以红绿为主色调，引领宾客进入了一个看似静止的奢华世界。
4. 浅色的古典装饰凸显了空间和灯光的感觉。

Dining Area
餐饮区

More Elegance in Dining Area
让就餐更高雅

The dining area is a place for dining and communication, where the environment, dining tables, passages, etc. form an integral area for dining. Apart from satisfying the dining requirements, the restaurant is supposed to boast certain artistic values, giving the guests more aesthetic enjoyment. Reasonable and comfortable dining environment can be realised by allocating and utilising the space efficiently. The size of passages, the arrangement of tables and chairs, the distance from kitchen, the circulation of guests and service staff, all of which are related to the rational design of the whole restaurant. In addition, restaurant requires appropriate decoration for creating elegant atmosphere. As unified dining tables in the restaurant occupy a large area, the awkward design for restaurant would be monotonous. Hence, the careful design on the interface around the tables is needed especially in the ceiling design and the use of lighting. As the decoration, the large fresco is the main element to highlight the neo-classical style, which is commonly used in a ballroom with a large area. In the ballroom where wedding ceremonies or other business events are held, elegant fresco can better enhance the solemn atmosphere of the space. And in a relatively small restaurant, it focuses more on creating a warm setting by employing original paintings, artistic works and the warm colour fabric, which are the common elements.

The big restaurant in Taleon Imperial Hotel is exquisite and stately. Fresco with little angels decorates the ceiling. The pilaster and carved paneling decorating the wall have the same palette of light blue as the fresco on the wall. The black fireplace in the middle is filled with classical flavor; scarlet carpet and chairs add heated atmosphere in which dining becomes elegant unconsciously.

餐饮空间是满足顾客用餐、交流的场所，周围的环境、餐桌、过道等组合成了一个整体的就餐区域，除了满足就餐需求之外，餐厅还应该有一定的艺术价值，让顾客有更多的审美享受。就餐的舒适合理可以通过空间的有效分配和利用实现，过道的大小、桌椅的摆放、与厨房的距离、顾客与工作人员的流动路线等都关乎着整个餐厅的合理性。除此之外，餐厅还需要适度的装饰营造高雅的氛围。餐厅中统一的餐桌占用了很大面积，如果设计不当会使餐厅显得单调乏味，因此需要在周围的界面上做精心处理，尤其是天花板的设计和灯光的使用。

大型的壁画装饰是突出新古典风格的重要元素，在面积较大的宴会厅中常用，这种类型的宴会厅常用作婚宴或商务宴请等活动，格调高雅的壁画能更好的提升空间庄严的气氛。而小型的餐厅更强调温馨的氛围，原创的油画、艺术品、暖色的织物是常用的元素。

塔里昂帝国酒店的大餐厅精致宏伟，天花板上装饰着小天使壁画。墙壁上带有壁柱和雕刻镶板，和墙上壁画同为清新的淡蓝色色调。中间的黑色壁炉充满了古典的气息，大红的地毯和椅子烘托出了热烈的气氛，在这样的氛围中，就餐也在不自觉中变得优雅。

Taleon Imperial Hotel

塔里昂帝国酒店

Location: Saint-Petersburg, Russia
Designer: Jean-Baptiste Michel Vallin de la Mothe
Photographer: Taleon Imperial Hotel
Area: 6,100m²

项目地点：俄罗斯，圣彼得堡
设计师：让-巴伯蒂斯特·米希尔·凡林德拉莫特
摄影师：塔里昂帝国酒店
项目面积：6,100平方米

Taleon Imperial Hotel is the only modern luxury hotel in Saint-Petersburg built in the 18th century. Each of the 89 rooms has its own individual architecture, exclusive design and appearance.

The hall is decorated by unique standing lamps made of crystal at the Baccarat factory in France commissioned for the mansion's former owners. In the restoration of the inlaid parquet extremely rare green oak was used, a material so rare that even the State Hermitage does not have any items made from it. Its 19th-century interiors are the ideal place for grand banquets and conferences. The ballroom has a traditional style, that of the magnificent Baroque of the 18th century, with double windows requisite for such rooms. Reliefs on the walls and ceiling, an ornate ceiling painting, a fireplace of white Italian marble, candelabra, and enormous mirrors – the hall's decorations have been preserved from the time of the Eliseevs. It is these mirrors that until 1917 shone at glittering balls, mascarades, nights of music, grand assemblies, and later, when it was a part of the building of the legendary House of Arts, poetry readings by Alexander Blok, Vladimir Mayakovsky, Osip Mandelstam, Nikolai Gumilev and other celebrated Russian poets

The spacious rooms are characterised by optimal use of space, harmony of design details and colors. Classical style of hotel interior combines the latest advances of modern technology with the grandeur and charm of the past centuries. The special feature of the Taleon Imperial Hotel is the use of ecological materials. Natural materials such as wood, malachite, marble, flax are used here. Different paintings and antiques made of bronze and marble are also used for rooms' decoration.

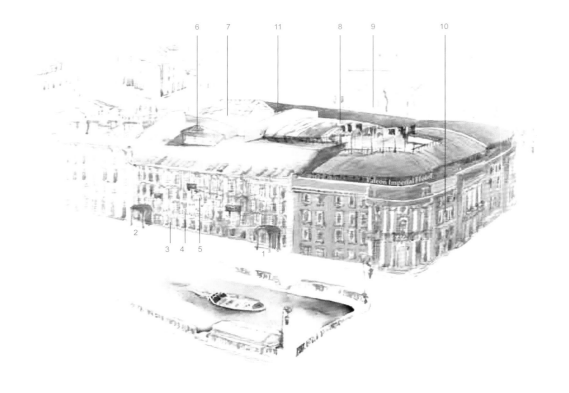

Hotel plan
1. Main entrance
2. Club entrance
3. Elieev bar
4. Taleon restaurant
5. Historical halls
6. Victoria restaurant
7. Atrium café
8. Taleon SPA
9. Griboedov restaurant
10. Gala enfilade and Imperial grand hall
11. Eliseev's banya

酒店平面图
1. 主入口
2. 俱乐部入口
3. 伊利芙酒吧
4. 塔里昂餐厅
5. 历史大厅
6. 维多利亚餐厅
7. 中庭咖啡厅
8. 塔里昂水疗馆
9. 格里博耶多夫餐厅
10. 节日厅和帝国大厅
11. 伊利芙桑拿

塔里昂帝国酒店是圣彼得堡18世纪所建造的唯一一家奢华酒店,它的89间客房都具有独立的建筑设计与外观。酒店大厅采用了独特的水晶落地灯(由宅邸的前业主委托法国巴卡拉工厂制造)进行装饰。镶木地板的修复选用了异常稀有的绿橡木,这种材料埃米塔日州甚至都没有出现过。酒店的19世纪风格室内设计十分适合举办大型宴会和会议。

宴会厅采用了传统风格,拥有18世纪的华丽巴洛克风情,每间房都采用了双层玻璃。墙壁和天花板上的浮雕、绚丽的天花板绘画、意大利白色大理石壁炉、枝形大烛台和大量镜子的使用让大厅显得富丽堂皇,保持了爱丽丝弗斯时期的装饰风格。这些镜子直到1917年都在闪耀的舞厅、假面舞会、音乐会、大型集会以及亚历山大·勃洛克、马雅可夫斯基、曼德尔斯塔姆、古米立夫和其他俄罗斯著名诗人的诗歌朗诵会上大放异彩。

宽敞的客房优化配置了空间,和谐地结合了细部设计和色彩。古典风格的酒店室内设计结合了最新的现代技术与古典的华丽风格和魅力。塔里昂帝国酒店的特色之一是它采用了生态材料。设计采用了木材、孔雀石、亚麻制品等天然材料。客房内部还采用了各种各样的绘画作品和青铜、大理石古董进行装饰。

1. Taleon Restaurant is one of the few places in St. Petersburg where the guests can dine in historical 19th century palatial interiors.
2. This two-room historical suite, which measures over 120 square metres, is the former bedroom of the prerevolutionary mistress of the palace, Varvara Eliseeva.
3. Luxury Studio Suites are one room apartments with ample space and feature king size beds.

1. 在圣彼得堡,塔里昂餐厅是为数不多的拥有19世纪豪华内饰的餐厅。
2. 双室历史套房总面积120多平方米,是宫殿前女主人瓦尔瓦拉·伊利西瓦的卧室。
3. 奢华工作室套房是一居室公寓,宽敞的空间内配有特大号双人床。

1. Legendary Premium suite, decorated with Corinthian columns, furnished with antiques, furniture, produced at the best European factories, and plush carpets, gracing the floor, leave no doubt that this is a room fit for royalty.
2. Executive suite designed in elegant pastel colours provide a wonderful view to the historical sights of the city.
3. One-room Superior Suite has a city view.
4. Superior room is spacious and cozy room with a city view or a view to the Atrium of the hotel.

1.传奇顶级套房装饰着科林斯柱，配有古董、出自欧洲最好的工厂的家具、奢华的地毯，具有皇室气息。
2.行政套房采用了优雅的粉蜡色设计，享有古城美景。
3.一室高级套房享有城市美景。
4.高级客房宽敞而舒适，享有城市或酒店中庭的景色。

Public Area
公共活动区

Traditional Meeting Room Nurtures High Technology
传统会议室孕育高科技

Holding large business conferences and providing high-quality business service are the requisite function in a high-end hotel. A perfect meeting room should be equipped with excellent audio-visual equipment, appropriate temperature and humidity, favorable ventilation device, etc., and the decor should be simple. Furthermore, due to the different demands, the meeting room should possess a certain degree of flexibility in order to improve the service efficiency. As much more requirements on neo-classical style decor, the style is usually simplified in the design of meeting room by using some abstract signs to present the neo-classical atmosphere. To achieve perfect effect, modern devices are indispensable. For ensuring the effect of the devices, the palette of meeting room should adopt soft light colours instead of bright colours with high saturation. The Wall and the board should be made by the non-reflective materials.

Eynsham Hall owns 53 meeting rooms with large windows ensuring the favorable ventilation and lighting. The interior is decked with wooden paneling, complying with the style of the entire hotel. Complicatedly carved fireplace inherits the elegance of the classical style, featuring the unified colour palette with walls. Although the interior is full of classical flavours, the devices in the meeting room are very advanced. Besides high quality audio and projection devices, each meeting room is connected with high-speed Wi-Fi. These modern devices in the meeting room do not make people feel abrupt, whose colours and forms echo the interior decor and combine subtly with the environment.

承办大型商务会议、为顾客提供高质量的商务服务是高档酒店必备的功能。一个完善的会议室需要有良好的视听设备、适合的温度和湿度、良好的通风条件、隔音设施等，装饰也应以简洁为主，另外，由于不同会议的要求，会议室还应具有一定的灵活性，以提高其使用效率。新古典风格对装饰要求较多，因此会议室的设计中常常会把这种风格加以简化，抽象出一些符号来表现新古典的氛围。为了达到完美的效果，现代化的设备必不可少，而为了保证这些设备的使用效果，会议室的色调应尽量采用柔和的浅色，而避免饱和度较高的鲜艳颜色。墙面、会议桌等也应用不易反光的材质。

艾恩汉姆会馆酒店有53间会议室，会议室带有宽大的窗户，保证了良好的通风和采光，室内采用了木制镶板的装饰，与酒店的整体风格相同，雕刻复杂的壁炉与墙面色调统一，有着古典风格的优雅。虽然装饰古典，但为会议室的设施却非常先进，除了高品质的音响、投影设备之外，每间会议室都有高速的Wi-Fi连接。这些现代设施在会议室中并不突兀，它们的色彩和形式都呼应了室内的装饰，巧妙地与环境融为了一体。

Eynsham Hall

艾恩汉姆会馆酒店

Location: Oxford, UK
Designer: Project Orange
Photographer: Richard Learoyd

项目地点：英国，牛津
设计师：橘子工程公司
摄影师：理查德·利亚洛依德

Situated on the outskirts of the city of Oxford in extensive grounds, Eynsham Hall is a magnificent countryside mansion. Completely remodelled in 1908 by Lady Mason, Eynsham Hall is wonderful example of the Jacobean Style very much in vogue at the turn of the twentieth century. With an enormous leaded window with views out into the extensive grounds, original feature oak panelling and magnificent carved feature fireplace, the meeting room in its existing state was quite fabulous. The brief, to transform the room into a fully functioning, state of the art boardroom with flexible lighting and audiovisual requirements though simple, was by no means an easy task.

An interconnecting door with the meeting room's neighbour the Gun Room bar, also introduced a further use as a potential private dining space.

A giant pop art inspired vortex rug sits beneath the table, also endlessly reflected in the table's mirrored base. Recent graduates from the Royal College of Art provide the framed artworks for the panelled walls. The tongue in cheek, burning logs above the fireplace add to the cheeky sense of humour, which pops up throughout. The original safe, presumably housing the Mason's family fortune has been retained as a quirky form of additional storage.

The first floor deluxe bedroom with a dramatic bay window and a rather fabulous original marble fireplace has been brought back to life with a mixture of old-fashioned luxury and quirky modern highlights; a traditional "knoll" two-seater sofa is upholstered in a vibrant, flock fabric and simple white, panelled wardrobe doors with a cheeky lime green lining. The bathroom is treated as a panelled box which acknowledges the fact that it is a new addition, but one that is bespoke and accommodates the padded headboard. Inside, the bathroom is tiled in dramatic black marble taking its clue from the fireplace. The smaller rooms also feature the white raised and fielded panelling, heavy cornicing and grey walls lit with contemporary lighting. Overall the furniture and colour scheme seek to create a comfortable and stylish interior that looks over its shoulder to the past but refreshing and modern at the same time.

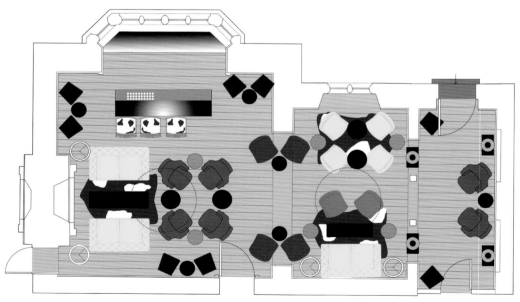

Bar plan 酒吧平面图

1. The main reception lounge retains many of its classic features.
2. The bar emphasises Eynsham Hall's traditional past as a fine country house.

1. 主会客厅保持了许多古典特色。
2. 酒吧凸显了艾恩汉姆会馆的乡村别墅背景。

艾恩汉姆会馆位于牛津城郊一处广阔的土地上,是一座宏伟的乡村庄园。会馆在1908年由梅森夫人进行了彻底翻新,是20世纪初詹姆士一世风格的典范。

巨大的铅制窗口能够看到外面广阔的庄园;橡木镶板和华丽的雕刻壁炉让会客室显得富丽堂皇。项目要求将房间改造成一个全功能、高科技会议室,配备灵活的灯光和音响设施。这个要求看似简单,实则不然。而连接会客室的军械库酒吧,也可以作为一个潜在的私人就餐空间。

桌子下方铺着一张流行艺术风格的漩涡图案地毯,在桌子的镜面底座上倒映出无数个映像。壁炉里燃烧的原木增添了房间中的幽默感,劈啪作响。原来珍藏梅森家族财富的保险室被改造成为独特的造型,用作额外的储藏室。

酒店一楼的奢华卧室带有一个夸张的飘窗和华丽的大理石壁炉,为居住生活带来了旧式奢华和现代感的混合效果;传统的双人沙发上表面采用了充满活力的植绒面料,而简洁的白色衣柜门则配有石灰绿色的衬线。浴室被打造成了一个镶板盒子,是新加的设施。浴室内夸张的黑色大理石从壁炉设计中获得了灵感。较小的房间采用了白色的浮雕镶板、厚重的飞檐和现代照明的灰色墙壁。所有家具和色彩搭配都旨在营造舒适时尚的室内设计,既体现了建筑的历史氛围,又增添了现代气息。

1. Guests can relax in the main lounge which retains its wooden floor and original stone fireplace, complemented by a new log fire burner and elegant, modern furnishings.

1. 宾客可以在酒廊中休息;酒廊保持了原有的木地板和石壁炉,并新配置了原木燃烧炉和优雅的现代装饰。

1. Developed around a Grade II listed, Jacobean-style mansion, Eynsham Hall has 53 meeting rooms.
2. Fireplace detail
3. All the meeting rooms benefit from natural daylight.

1. 艾恩汉姆会馆围绕着詹姆士一世风格进行装饰，共有53间会客厅。
2. 壁炉细部
3. 所有会客室都受益于自然光。

Meeting room elevations 会议室立面图

075

1. Traditional accommodation is period charm and elegance.
2. This fine Jacobean-style mansion has a long tradition of hospitality.

1. 传统客房拥有古典的优雅魅力。
2. 这家詹姆士一世风格的会馆有很长的接待历史。

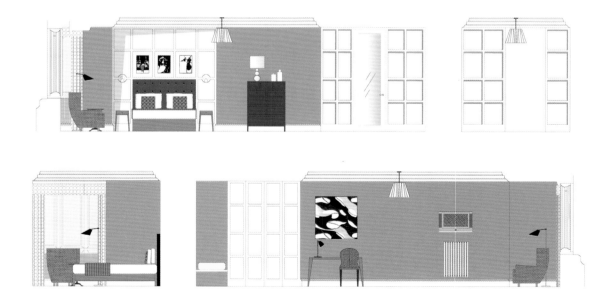

Standard bedroom elevations 标准客房立面图

Delux bedroom plan

1. Entrance
2. Bathroom
3. Bedroom

豪华客房平面图

1. 入口
2. 浴室
3. 卧室

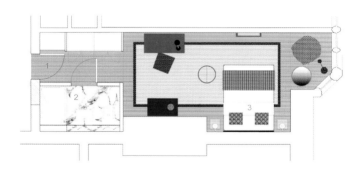

Standard bedroom plan

1. Terrace
2. Bedroom
3. bathroom

标准客房平面图

1. 露台
2. 卧室
3. 浴室

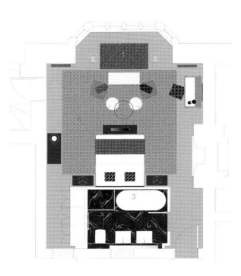

Chapter Three

INTERFACE DESIGN IN NEO-CLASSICAL HOTELS
新古典酒店的界面设计

Chapter Three
INTERFACE DESIGN IN NEO-CLASSICAL HOTELS
新古典酒店的界面设计

A space is enclosed with ceiling, walls and ground. These three interfaces with large areas are the main parts showcasing the space style. The interface design should firstly serve the overall style of space and present the characteristic by mouldings, colours and materials. Moreover, it should create the atmosphere the space requires, which can reasonably blend with other elements in the space and express the theme of design. In order to present the neo-classical atmosphere, the interfaces in the neo-classical style hotel have complicated mouldings, especially on the ceiling and walls. They are presented by carving, inlaying and various decorations, which add more difficulties to the design.

The structure of the hotel ceiling is complicated and confined greatly to its design. In addition to the consideration on the attractive appearance, lighting, air-condition, air return and other issues should be also taken into consideration. In the neo-classical style that pursues preciseness and

一个空间是由顶面、墙面和地面三个界面围合而成，这三个界面面积大，是体现空间整体风格的重要部分。界面的设计首先要服务于空间的整体风格，在造型、色彩、材料上体现空间的风格特点，另外还要营造出空间所需要的气氛，与空间内其他的元素有机结合，表达出设计的主题。为了体现古典气氛，新古典风格酒店的界面一般造型复杂，尤其是顶面和墙面。雕刻、镶嵌以及各种装饰品都有所体现，这也为设计增加了一定难度。

酒店的顶面结构繁复，设计也有很大制约，除了考虑造型美观的要求之外，还要考虑灯具、空调系统、回风孔等设备要求。而追求严谨和华丽新古典风格中，顶面的设计更为复杂。在欧洲一

The beautiful neo-baroque ceiling paintings recall the glamour of days gone by.

漂亮的新巴洛克风格棚顶壁画让人想起了往日的辉煌。

Chapter Three

INTERFACE DESIGN IN NEO-CLASSICAL HOTELS

luxury, the ceiling design is more complicated. In some European hotels which are renovated from historic buildings, magnificent vaults are usually the most gorgeous scenery in the hotel. These vaults are often decked with frescos carved by famous artists. With hundreds of years of history, these frescos are generally bright-coloured and full of meaning, and usually adopted in the relatively large lobby and ballroom. In modern neo-classical hotels, though such frescos are not frequently used as before, elaborate art deco is still indispensable. The carved ceiling divided by beams can perfectly match up with wooden paneling on the wall. Colourful plaster ceiling with delicate patterns and the simple-moulded plaster, in particular, is also a popular way for the neo-classical style

些由古建筑翻修而成的酒店中，华丽的拱顶常常是酒店中最为绚丽的风景，这些拱顶常常装饰着名家雕刻的壁画，有着几百年的历史，壁画一般色彩艳丽并且富有寓意，常用在酒店中面积较大的大堂或者宴会厅中。在现代的新古典酒店中，虽然这种壁画不再常用，但精心的装饰也必不可少。用木梁分隔并带有雕刻的顶面可以很好地配合墙面的木制镶板，彩色的石膏天棚雕刻精美的图案也是新古典风格中常用的形式，尤其是造型简洁的石膏装饰最为常见。当然，无论哪种天棚，华丽的水晶灯是新古典空间里永恒的搭配。

The magnificent Grand Restaurant in neo-baroque style, with paintings on the walls and ceiling dating from 1901, is simply enchanting.

宏伟的大餐厅采用了新巴洛克风格，墙壁和天花板上的壁画源于1901年，极富魅力。

presentation.

Walls are the vertical interface, which is the important element to divide the space. Also, it plays a role of connecting the ceiling and the ground, exerting a great influence visually. At the same time, walls can preserve the temperature, insulate the noise and heat, etc. Therefore, certain consideration must be given to both the appearance and function on the selection of materials and the moulding design. The wall design for the neo-classical hotel adopts varied methods depending on the different function areas. The lobby covers a huge area and is expected to be magnificent and gorgeous, hence, natural stone, in particular, the marble is the appropriate material. However, in guestroom and small public areas, with delicate carved patterns, wooden paneling is more popular. In the modern hotel design, convenient and varied wallpapers are welcomed by more and more people, especially in hotel guestrooms. To highlight the neo-classical style, elegant and colourful wallpapers are usually adopted, and oriental silk wallpapers are used in some senior suites.

The ground should bear the burden of the furniture and the circulation of guests, so the parts revealed on the ground are relatively less. However, it also influences the interior design and has higher demands on abrasive resistance, waterproofing and fire proofing. In public areas of the hotel, such as lobby, restaurant and ballroom, where guests frequently shuttle, the marble ground is generally adopted. The stone not only has perfect abrasive resistance and decorative effect, but is easy to clean. In the guestroom, the more private space, wooden floor is the best choice for the neo-classical style presentation. Wooden floor is natural and plain, which can create comfortable and warm atmosphere and echo the theme of the neo-classical environment. Whether the marble ground in public areas or the wooden floor in guestrooms, they are necessarily matched up with carpets. The printed carpet coordinating with the design theme is an indispensable element of the neo-classical style design.

墙面是垂直的界面，是划分空间的重要元素，也起着连接顶面和地面的作用，在视觉上对人的影响很大。同时，墙面还担负着保温、隔音、隔热等作用，所以，在材料的选择和造型设计上需要同时兼顾美观和功能的要求。新古典风格酒店的墙面处理根据功能区的不同会采用不同的方式。大堂面积较大，且要求有宏伟华丽的气势，因此，天然的石材尤其是大理石是合适的材料。而在客房和小型的公共空间中，木制镶板更为常见，并且镶板上一般都带有精致的雕刻。在现代的酒店设计中，使用方便并且花样繁多的壁纸越来越受到人们的喜爱，尤其是在酒店的客房中，为了突出新古典风格，一般采用表面雅致绚丽的壁纸，一些高档套房还会采用来自东方的丝绸壁纸。

地面上需要承担家具和人的流动等，所以显露出的部分相对较少，但也直接影响着室内的装饰效果，并且对耐磨程度、防水、防火等要求较高。在酒店的公共区域，如大堂、餐厅、宴会厅等处，人流往来频繁，一般采用大理石装饰地面，石材不仅有很强的耐磨性和装饰性，而且也更易于清洁。而在私密性较强的客房空间，木制地板是新古典风格的最佳选择。木制地板自然朴实、能创造出舒适温暖的氛围，与新古典环境也相得益彰。而无论是公共区域的大理石地面还是客房中的木制地面，地毯都是必备的搭配，与设计主题相协调的印花地毯在新古典风格的设计中几乎是不可或缺的元素。

Ceiling Design
顶面设计

Gorgeous Ceiling
天花板上的绚丽

As hotel has many function areas, the ceiling design has varied features according to different spaces. In the neo-classical style, fresco is a frequently-used element in the public area. The content of fresco includes angels, saints, etc. which is usually related to some allegories. Around ceiling are usually carvings with complicated flora patterns or geometric figures. The large fresco and complicated carvings are apply to spacious public area for presenting magnificent momentum. In small guestrooms, plaster and wooden ceilings are more common. White plaster ceiling is carved complicatedly with gilt edges, while wooden ceiling is usually latticed or regular geometric figures, whose colours echo the design of ground. In the bathroom or spa, ceilings are designed in a more modern and direct way.

Grand Hotel Kronenhof is a neo-baroque style hotel. This style is presented incisively and vividly by the hotel ceiling design. Once stepping into the hotel, guests will see the florid fresco above the reception. Gold carvings around the fresco make the lobby magnificent. The ceilings of the lobby and restaurant are also decked with large frescos. Similarly, the one in the restaurant has a history dating back to 1901. Apparently, these ceilings have baroque features: they are colourful, apt to be carved into portraits, leaves, shells and waves making the illusion as if in the heaven. In guestrooms and small restaurants, plaster ceilings divided by beams are adopted. They are simple but full of classical elegance.

酒店的功能区很多，顶面的设计也根据空间不同的功能有不同的特点。在新古典风格中，壁画是大面积的公共区顶面常用的元素。壁画的内容一般是天使、圣经人物等，通常带有一定的寓意，而在顶面四周一般雕刻更为复杂的花草形状或几何图形。这种大型的壁画和复杂的雕刻适用于宽敞高大的公共区域，以体现恢宏的气势。在小面积的房间，石膏天棚和木制天棚更常见，白色的石膏天棚上可以做复杂雕刻，边界镶嵌金边，而木制棚顶一般设计为网格状，或有规律的几何图形，并且，棚顶的图案色彩会与地面的设计有所呼应。而在浴室或SPA等特殊区域，一般会直接采用更为现代的形式。

科隆霍夫大酒店是一家新巴洛克风格的酒店，这种风格在酒店顶面的设计上得到了淋漓尽致的体现。一入酒店大门，就会看到接待处上方绚丽的壁画，周围是网格状的图案，四周金色的雕刻使门厅异常富丽。在大堂和大餐厅的棚顶，同样是大型的壁画装饰，餐厅的棚顶壁画甚至可以追溯到1901年。这些顶棚的设计带有明显的巴洛克特征：色彩斑斓、设计富有雕塑性、造型包括人像、树叶、贝壳、卷轴等，并且给人一种如苍穹般的错觉。而客房和小型的餐厅则用了石膏天棚和木梁分隔的顶面，简洁而不失古典的雅致。

Grand Hotel Kronenhof

科隆霍夫大酒店

Location: St. Moritz, Switzerland
Designer: Justus Dahinden, Rolf Som
Photographer: Grand Hotel Kronenhof
Area: 8,000m²

项目地点：瑞士，圣莫里茨
设计师：贾斯特斯·达辛登；洛尔夫·索姆
摄影师：科隆霍夫大酒店
项目面积：8,000平方米

The 5-star superior Grand Hotel Kronenhof in Pontresina is one of the most architecturally significant Grand Hotels in the Alps since the 19th century. Many of the luxurious rooms and suites are in décor of a patrician dwelling, which comprise elegant, lavishly furnished rooms pampered with the latest technology to provide a luxurious home away from home. The historic and awarded Restaurant Kronenstübli, with its traditional Grisons wooden furnitures and dark pine panels, delights all gourmet connoisseurs with an exquisite cuisine.

Wellness in the Grand Hotel Kronenhof means direct contact with forests, water and mountains from the moment of arrival. Everywhere, be it in the treatment rooms, the gym or the guestrooms, tired eyes are comforted with the sight of Swiss pine, larch and the magnificence of the Roseg glacier. The spa, an oasis stretching over 2,000 square metres, is specially designed to bring natural freshness inside and ensures the hotel is first choice for all those wanting to stay naturally healthy.

The Grand Hotel Kronenhof offers 9 charming suites: 6 suites are located in the south wing of the hotel, 3 suites are located in the west "Ganzoni" wing. Every suite is individually designed comprising rich, comfortable accommodation with an en-suite luxury bathroom, some feature walk-in wardrobes. Southern-facing suites offer a unique view of the Roseg glacier. All rooms and suites facing south or southwest have inspiring environments, with splendid panoramic views of mountains and Corviglia and Roseg Valley glaciers. Ranging in size from 35 m² to 75 m², Junior suites and suites feature luxurious equipped bathrooms in marble and granite, some with walk-in wardrobes. Living and sleeping areas of the 9 suites, 22 Deluxe Junior suites and 15 Junior Comfort suites present their own individual style and attention to detail.

五星级奢华酒店科隆霍夫大酒店是19世纪以来阿尔卑斯地区最宏伟的酒店之一。许多奢华的客房和套房都采用了贵族宅邸的装饰。装饰优雅的房间搭配着最先进的技术，打造了第二个奢华的家园。历史悠久的获奖餐厅科隆斯塔布里采用了传统格里松斯木家具和深色松木板，以精致的美食愉悦着食客。

科隆霍夫大酒店的健康生活直接与森林、水流以及山川相连。无论是在理疗室、健身房，还是客房，疲惫的双眼都能从瑞士松木、落叶松和冰川中获得休憩。延伸2,000平方米的水疗中心是一片绿洲，专门提供天然的新鲜体验，是寻求健康生活的宾客的首选。

科隆霍夫大酒店拥有9间魅力十足的套房：6间位于酒店南翼，其余的3间位于西翼。每间套房的设计都与众不同，提供丰富、舒适的住宿环境。朝南的套房可以观赏到罗斯格冰川的独特景色。所有朝南或西南的客房和套房都享有山川、考尔维利亚和罗斯格山谷冰川的壮丽景色。普通套房和套房面积从35到75平方米不一，以奢华的大理石和花岗岩浴室和步入式衣柜为特色。9间套房、22间奢华小套房和15间普通舒适套房的起居区和睡眠区都拥有独特的风格和细节设计。

1. The opulent lobby was created by Otto Haberer in the early 1900s, and has been restored to its former glory.
2. The hotel reception in neo baroque style

1. 豪华的酒店大堂由奥托·哈波尔设计于20世纪初，经过翻修重现了往日的辉煌。
2. 新巴洛克风格的酒店前

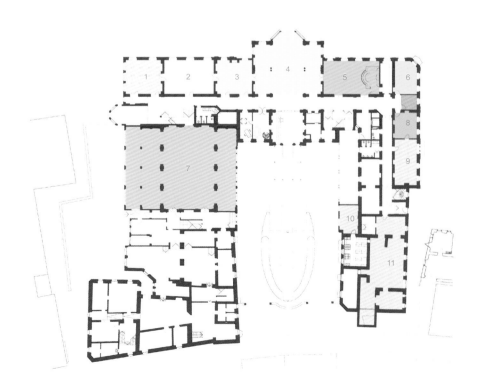

Floor plan
1. Salon rose
2. Salon bleu
3. Kamin-zimmer
4. Lobby
5. Bar
6. Salon julier
7. Grand resto
8. Billard
9. Trais fluors
10. Salon languard
11. Kegelbahn

平面图
1. 玫瑰沙龙
2. 蓝色沙龙
3. 卡明奇默厅
4. 大堂
5. 酒吧
6. 朱莉沙龙
7. 大餐厅
8. 台球室
9. 三味厅
10. 蓝嘉德沙龙
11. 撞球室

1. Cosiness and the feeling of great comfort are the characteristics of this wood panelled room.
2. The Kronenhof Bar & Lounge has flower shape tables.
3. Grand lobby

1. 镶木装饰的房间令人感到舒适和放松。
2. 酒吧兼休息室有花朵形状的桌子。
3. 宏伟的大堂

1. The Fireplace Lounge is the natural extension to the Hall. With its sumptuous leather armchairs it provides a comfortable and cosy ambiance while listening to the crackling sound of the fireplace on cooler nights

2. The newly redesigned "Bellaval Suite" surprises with luxurious comfort and meticulous attention to detail. Numerous large windows contribute to this light-flooded room and offer breathtaking views of Glacier Roseg and St. Moritz mountain range. The exquisitely finished décor, the elegant furnishings and warm earth tone colours create a comfortable and pleasing ambiance.

1. 壁炉酒廊是礼堂的自然延伸。豪华的皮革扶手椅为酒廊营造出舒适的氛围。寒冷的夜晚，壁炉中燃烧的木材会劈啪作响。
2. 重新设计的贝拉沃套房奢华舒适，每个细节都至真至美。宽敞的窗户让房间内洒满了阳光，并展现了罗斯格山谷冰川和圣莫里兹山脉的壮丽景色。精致的装饰、优雅的家具和温暖的大地色调营造出舒适愉快的氛围。

3. Valentine suite is tastefully designed and offers a separate living- and bedroom area as well as an open design bathroom in grey marble. Wood carved from Engadine mountain spruce creates a delightful mix of warmth and style.
4. Living area invite to relax
5. Wood carved from Engadine mountain spruce creates a delightful mix of warmth and style.

3. 情人节套房精心打造了独立的起居区和卧室区，并且在浴室采用了灰色大理石设计；来自恩加丁山的云杉木营造了温馨愉悦的风格。
4. 起居区让人感到放松。
5. 来自恩加丁山的云杉木为精致的套房带来了传统的舒适和独特的风格。

093

Wall Design
墙面设计

Surrounded by Art
被艺术包围

Vertical wall usually makes great visual impact, and it is an important part in the hotel design. Besides its own design style taken into account, its relationships with windows, doors and the relative ceiling should be considered as well. The wall can be decoratedwith stones, frescos and wallpapers in the neo-classical style. Frescos and stones are usually used in public areas. Especially in the hotels with long history, frescos are prevailed. Generally, these frescos opt for classical themes, echoing the frescos on the ceiling and reflecting a strong sense of history. Wooden walls are easy to install and apt to be carved, which are applied widely in neo-classical hotels. The walls made by wooden paneling are divided into several areas in a certain proportion, creating a harmonious and symmetrical effect. White paneling with gilded edges is the most common match. Besides, the decorations on walls are inevitable, as well as tapestries and original paintings are the best selections.

Villa Le Rose is a hotel renovated from a private residence. The guestrooms, the restaurants, the ballrooms, even the restrooms of hotel are decked with a large area of frescos. Sprayed with strong or delicate shades of colours, the subjects presented on the frescos involve plants, characters and animals, which are integrated with the other frescos on the ceiling into a harmonious whole, where guests feel as if they were exposed to the scenes on the frescos: the restaurant on the ground floor is like that of the country of 18th century; the King Premiere Suite resembles the medieval palace; even in the bathrooms, the strong artistic atmosphere can be sensed. Gorgeous textiles and soft lighting add special scenery to these arty worlds. Here, walls play a leading role in design and every detail is expected to be compatible with it, telling a parade of stories.

垂直的墙面对人的视觉冲击力很大，是酒店设计中重要的部分，除了要考虑自身的设计风格之外，还要考虑和界面中的窗户、门以及与之连接的天棚之间的关系。新古典风格的墙面处理一般有石材、壁画、木制镶板和墙纸这几种方式。壁画和石材一般应用在公共空间，尤其在一些历史悠久的酒店中，壁画的应用很多。这些壁画一般选择古典题材，和天棚壁画相呼应，有很强的历史感。木制墙面安装方便，且易于雕刻，在新古典风格的酒店中应用很广泛。这种木制镶板的墙面会被雕刻分割成带有一定比例的区域，产生和谐对称的效果，白色镶板和镶嵌的金边是常见的搭配。除此之外，墙面的装饰物也不能缺少，挂毯、原创油画是最佳的选择。

玫瑰别墅酒店是一家由私人住宅改造成的酒店。酒店中的客房、餐厅、宴会厅，甚至卫生间都装饰了大面积的墙面壁画。壁画的内容有花草、人物、动物，色彩或浓烈或淡雅，并且与天棚上的壁画浑然一体，置身其中，仿佛被壁画中的场景包围：一层的餐厅如同18世纪的乡村；国王套房如同中世纪的宫殿；即使在浴室也同样能感受到酒店浓浓的艺术气息，而华美的织物和柔和的灯光，更为这些艺术的世界增加了别样的风景。在这里，墙面是设计中的主体，室内的每个细节都配合着墙上的壁画，讲述着一个个故事。

Villa Le Rose

玫瑰别墅酒店

Location: Florence, Italy
Designer: Leonardo and Beatrice Ferragamo
Photographer: Villa Le Rose
Area: 4000m²

项目地点：意大利，佛罗伦萨
设计师：李奥纳多·菲拉格慕；碧翠丝·菲拉格慕
摄影师：玫瑰别墅酒店
面积：4,000平方米

Villa Le Rose stands at the end of a great driveway of ancient cypresses. It is built around a courtyard and is surrounded by gardens and olive groves.

During the course of the refurbishment, the villa was fitted with all the modern comforts and practicalities while maintaining and enhancing its historical features. Through the formal entrance, you enter into " the cortile" which is decorated with large terracotta vases and jasmine and no less than a gorgeous Della Robbia ceramic Madonna and Child which is worthy of a place in Florence's great museum Gli Uffizi. The loggia, perfect for summer days (as it is always quite cool) is decorated with wrought iron chairs and couches.

There is a sumptuous ballroom decorated with frescoes, a massive crystal chandelier and a grand piano. The villa also has a formal dining room which can seat up to 22 people, and a large living room which can comfortably seat 14. For movie watching, satellite TV or simply relaxing, there is a library.

Next to the Villa is the Cedar Garden, a sweeping lawn enclosed by high hedges and ancient trees, embellished by antique rose bushes and pietra serena pedestals bearing potted lemon trees, and a 300 year old Lebanese cedar.

The Villa is built on several floors according to the layout typical of aristocratic villas. On the ground floor is the Sala delle Feste, a ballroom which is a fine example of 18th century country-residential architecture, leads onto by spacious living room with frescoed walls, and a welcoming study and a comfortable dining room.

The "Baldacchino Suite", also on ground level, is distinctive for the richness of this furnishings and decorations, with a great canopied bed and an entirely frescoed bathroom with fireplace.

On the upper floor are five suites, each of them embellished with original furnishings, decorated ceilings and large windows opening onto a striking unforgettable landscape.

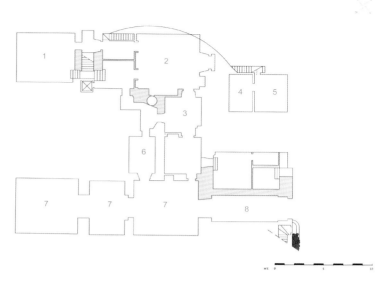

Ground floor plan
1. Storage
2. Kitchen
3. Breakfast room
4. Elevator shaft
5. Cellar
6. Utility room
7. Garage
8. Vehicular entrance

一层平面图
1. 储藏室
2. 厨房
3. 早餐室
4. 电梯井
5. 地窖
6. 杂物间
7. 车库
8. 车辆入口

1. The spacious living room has frescoed walls.
2. On the ground floor, a magnificent 18th century ballroom complete with grand piano and impressive chandelier.
3. First floor landing, has the same tone with public area.
4. There is a dressing table in Green Bedroom

1. 宽敞的客厅采用了壁画墙壁。
2. 一楼宏伟的18世纪风格宴会厅内装饰着大钢琴和令人印象深刻的吊灯。
3. 二楼楼梯平台与公共区域采用了同样的风格。
4. 绿色客房的梳妆台

1. The entrance is located under the beautiful arched loggia furnished with wrought iron chairs and couches and once inside; the hall is adorned with antique mirrors, a large antique wooden table and a decorative bronze hunting dog collection.
2. Informal sitting room which is also known as the TV room with full satellite TV, DVD player and a wonderful collection of sailboat models; it's also a beautiful library.
3. Landing entry to Guest Suite on first floor

Hotel plan
1. Bedroom
2. Bathroom
3. Studio
4. Sitting room
5. Entrance
6. Ballroom
7. Dining room
8. Hall

酒店平面图
1. 卧室
2. 浴室
3. 工作室
4. 客厅
5. 入口
6. 宴会厅
7. 餐厅
8. 大厅

1. 大门设在华丽的拱顶游廊之下，游廊里摆设着熟铁座椅和躺椅；酒店大厅内装饰着古董镜子、巨大的古董木桌和铜质的装饰性猎狗收藏品。
2. 日常起居室又称电视房，配有卫星电视、DVD播放机和大量帆船模型；它还是一个美丽的图书室。
3. 二楼套房的平台入口

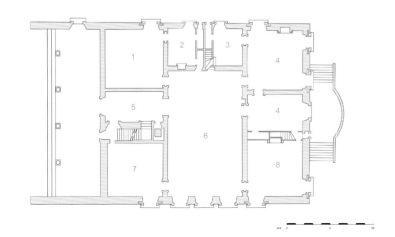

玫瑰别墅酒店位于一条两侧都是古柏的大道尽头。酒店围绕着一座庭院而建，四周环绕着花园和橄榄树林。

在翻修过程中，酒店配备了舒适的现代化设施，同时也注重保留了其独特的历史韵味。

走进大门，你将看到由大陶罐和茉莉花点缀的内院。院子里的陶瓷圣母圣婴像可以与佛罗伦萨著名的乌菲兹博物馆中的藏品相媲美。适宜避暑的凉廊内装饰着熟铁座椅和躺椅。

酒店内的豪华宴会厅里装饰着壁画、巨型水晶吊灯和一台大钢琴。酒店的正餐厅可容纳22人，大型会客厅则可容纳14人。客人还可以在酒店的图书室里看电影、电视或是休闲放松。

别墅旁的雪松花园的大草坪四周环绕着高高的树篱和古树，其间还点缀着古老的玫瑰丛和种在塞因纳式花坛里的柠檬树。当然，最显眼的当属一棵拥有300年历史的黎巴嫩雪松。

与经典的贵族别墅相同，玫瑰别墅分为几层。一楼是菲斯特大厅——一间经典的18世纪乡村建筑风格大厅。与大厅相连的是墙面上绘有壁画的会客室、书房和舒适的餐厅。

同样位于一楼的巴尔达奇诺套房，以其华美的家具和装饰而著称。套房配有宽大的四柱床，绘满壁画的浴室里还有一个壁炉。

别墅上层是五间各具特色的套房，它们都配有独特的家具和装饰吊顶。从套房的大窗户可以看到醉人的美景。

1. The dining room has gorgeous walls.
2. King-Bedroom located in ground floor.

1. 餐厅的墙面装饰异常壮观。
2. 国王卧室位于酒店一层。

1. Green Bedroom has king size bed.
2. The charming Le Rose suite which comprises of twin beds put together to make a king size bed; a comfortable seating area with two sofas and four very large windows offer fine views of Florence's beautiful rolling hills.
3. The Oriental Bedroom is the smallest bedroom but it is still very charming.
4. A beige room with a king size bed, being a corner suite this room also boasts views overlooking the Cyprus lined drive leading up to the villa.

1. 绿色套房的特大双人床
2. 迷人的玫瑰套房的两张床可拼为一张特大的双人床；舒适的休息区配有两张沙发，四扇大窗户展示了佛罗伦萨起伏的山峰。
3. 东方客房是最小的客房，但是依然十分迷人。
4. 配有特大双人床的浅黄色房间，由于它设在建筑的一角，所以可以俯瞰进入酒店的车道及其两旁的美景。

Floor Design
地面设计

Beautiful Views on the Floor
低着头欣赏

The floor design in the hotel should consider the following aspects: the protection against floor slab, the assurance to usage conditions and the effect of decor. To be specific, the materials for the floor should be anti-abrasive, anti-corrosive, waterproof, fireproof, cleanable, heatproof and insulated. At the same time, the materials should be neat and smooth in accordance with the decor style of the whole space. Marbles and wooden floors are common materials for neo-classical style presentation. Colourful marble floors with geometric figures and monochrome marble floors are chosen frequently. The wooden floor has a long history in the decor of neo-classical style. As early as in the 17th century, the wooden parquet floor was popular in the Europe. Timber with natural veins and texture is beautiful and comfortable, so it is preferable to be used in guestroom. Besides, carpets are widely adopted in some neo-classical hotels. In the lobby, carpets spread from the reception area to the lounge, then to the staircase, assuring the integrity as well as assisting in guiding the guests visually.

The Dorchester is a hotel with a long history. During its renovation, many luxurious materials are used. The ground of the lobby is paved with marbles with natural veins, matching with marble columns around. American walnut floors are adopted in most of the guestrooms. With plush texture, natural parabola veins on timber form beautiful patterns. Meanwhile, as its fixed size, ease of processing and high corrosion resistance, it is considered to be the suitable materials for the ground. However, as it is a kind of rare timber, walnut is costly and not largely used in the hotel.

酒店地面的设计要考虑几个方面的内容：对楼板的保护、使用条件的保证，以及装饰作用。具体来说，地面材料要保证具有必要的耐磨度、耐腐蚀、防水防火、易清洁、隔热保温等，同时表面要平整光滑，并且装饰风格与整体空间一致。大理石地面和木制地板是新古典风格常用的材料。彩色的大理石拼接成几何形状或黑白大理石的间隔交错在酒店大堂中是常见的选择。而木制地板在古典风格的装饰中更是有着悠久的历史，早在17世纪，木制拼花地板在欧洲就已经很流行。木材的天然纹理和质感既美观又舒适，非常适合酒店的客房中使用。另外，一些新古典风格的酒店中也常用大面积的地毯装饰地面，大堂中，地毯从接待处延伸到休息区，再到楼梯，保证设计整体性的同时也起到了视觉上的导向作用。

多尔切斯特酒店是一家历史悠久的酒店，在翻新过程中应用了许多豪华的材料。大堂的地面是天然纹理的大理石，配合周围的大理石柱。卧室中大部分用了美国胡桃木地板，这种木材有着天然的抛物线花纹，能形成美丽的图案，有着豪华的质感。同时它的尺寸稳定、易于加工、并且抗腐蚀力强，很适合用做地面材料，同时，胡桃木也是一种珍贵的木材，价格昂贵，因此很少大面积地使用。

The Dorchester

多尔切斯特酒店

Location: London, UK
Designer: Champalimaud
Photographer: The Dorchester
Area: 6,000m²

项目地点：英国，伦敦
设计师：查帕里芒德公司
摄影师：多尔切斯特酒店
面积：6,000平方米

Champalimaud completed the renovation of one of London's most prestigious collection of suites. Overlooking Hyde Park on Park Lane, The Audley, Terrace and Harlequin Suites, collectively known in the hotel as the "Roof Suites" for their lofty position on the top floor of the hotel, have long served as pied-a-terre for visiting celebrities and dignitaries. Theses Suites represent a contemporary glamour uncontested by any other suites in the city. Discretely reminiscent of the Hollywood glamour long associated with the hotel, the design of these suites emulates the style in which The Dorchester was originally built in the 1930's while remaining fresh, modern and luxurious. Luxurious materials sourced from around the world such as pleated silk walls, soft leathers and shagreen, exotic woods, beaded glass wallpaper, black moonstone and Calacatta Oro marble as well as bespoke furniture designed specially for these suites, create a stunning visual and tactile experience.

THE HARLEQUIN – The largest of the three rooftop suites. The Harlequin is made up of a master bedroom with an ensuite bathroom and dressing room, a dining room, living room, bar and large outside terrace overlooking Hyde Park. The rooms are arranged enfilade with American walnut floors throughout, giving the suite an elegant, stately flow.

THE AUDLEY – Pearls, dove grays, buttery creams, and sumptuously textured materials are characteristics of the Audley Suite. Black shagreen and crème coloured leather line the entry hall walls and the window in the stone floored foyer provides exposure to natural light and welcoming airiness.

TERRACE SUITE – The captivating features of the Terrace Suite that create an elegant sanctuary include a bathroom that opens directly onto the terrace; a master bedroom that extends into a private conservatory. The suite is served by a dedicated lift from the 8th floor that delivers guests directly to the private entry vestibule. Sumptuous leather-paneled walls in the inner hall offset the more delicate pearl-coloured Venetian plaster living room and pale pink Venetian plaster dining room. French walnut herringbone floors are throughout.

1. The Promenade is a series of rich, warm, intimate spaces culminating in a stunning, oval leather bar at the end of the room.
2. The Dorchester's new ballroom has opened after an extensive renovation. The original interiors by Alberto Pinto have been maintained, restored and enhanced. The major reworking however was carried out to the staircase making it grander and much more welcoming with Art-Deco inspired bronze finish balustrade.
3. The Gold Room, previously separate spaces, has been joined into one room to provide a more fluid pre-function space.
4. The Splendid Wedgwood style Orchid Room can be adapted from an intimate cocktail party to a formal conference.

1. 长廊由一系列华丽温馨的私密空间组成，最里面是一个椭圆形皮革吧台。
2. 多尔切斯特酒店的新宴会厅经过了大面积翻新；由埃尔伯托·平托所设计的室内装饰得到了保留、整修和改进；主要改造工作在于楼梯的拓宽以及装配装饰艺术风格的铜质栏杆。
3. 金色厅由一些独立的空间组合而成，提供了更流畅的准备空间。
4. 兰花厅以伟吉伍德陶器风格为特色，适合为正式会议举办小型鸡尾酒会。

Floor plan

1. Orchid
2. Orangery
3. Ballroom
4. Holford
5. The grill
6. Gold room
7. Silver room
8. Crush hall
9. Opal suite
10. Promenade
11. Reception
12. Park suite left

酒店平面图

1. 兰花厅
2. 橘园
3. 宴会厅
4. 霍尔福特厅
5. 烧烤餐厅
6. 金色厅
7. 银色厅
8. 迷恋大厅
9. 猫眼石套房
10. 长廊
11. 前台
12. 左侧公园套房

查帕里芒德公司完成了伦敦最具声望的酒店套房的翻修工作。俯瞰海德公园小路的奥德利套房、露台套房和丑角套房并称为酒店顶楼最著名的"屋顶套房",曾是诸多达官贵人的临时住所。这些套房展示了无与伦比的现代化奢华品味。

酒店套房的设计保留了部分其原有的好莱坞风格,同时效仿了20世纪30年代多尔切斯特酒店初建时的风格,显得清新、现代而奢华。

套房设计采用了包括褶皱丝面壁纸、软皮料和鲨革、异国木料、玻璃珠墙纸、黑月长石以及卡拉考特·奥罗大理石等来自世界各地的奢华材料,并配有定制的家具,营造出绝妙的视觉和触觉体验。

丑角套房——丑角套房是三间屋顶套房中最大的。丑角套房由配有浴室和化妆间的主卧、餐厅、客厅、吧台以及俯瞰海德公园的露天平台组成。套房铺设着美国胡桃木地板,显得优雅、庄重、流畅。

奥德利套房——珍珠色、鸽灰色、奶油色以及奢华的材质是奥德利套房的特色。门厅的墙上采用了褐色鲨革和奶油色皮革线进行装饰,而石铺门廊的窗户则为室内提供了充足的自然光线和新鲜的空气。

露台套房——露台套房的独特之处在于浴室直通露台以及主卧一直延伸到私人温室花房。客人可以从八楼的电梯直达套房的私人前厅。内厅奢华的皮革墙壁与客厅的珍珠色威尼斯石膏墙以及餐厅的浅粉色威尼斯石膏墙交相辉映。套房全部铺设着法国胡桃木人字纹地板。

1. The splendid wood-paneled sitting room provides an atmosphere of quality and tradition.
2. Situated above the front entrance of the hotel, Belgravia Suite is characterised by a beautiful four-poster bed.
3. Stanhope Suite bedroom is king or twin bedded, and the sitting room, much like a study.
4. Superior King Room is more spacious than the single rooms and affords the guest every luxury

1. 木质镶板会客厅显得高雅、传统。
2. 位于酒店大门上方的贝尔格莱威亚套房以华丽的四柱床为特色。
3. 马车套房的卧室配有一张特大双人床或两张单人床,它的客厅更像一间书房。
4. 豪华国王客房比单人客房更加宽敞,为宾客提供了更为奢华的体验。

Chapter Four

FURNITURE IN NEO-CLASSICAL HOTELS

新古典酒店的家具

Chapter Four
FURNITURE IN NEO-CLASSICAL HOTELS
新古典酒店的家具

The Furniture is a requisite in the hotel, and a main part in the hotel design, exerting a strong impact on the interior environment. The neo-classical style furniture is rooted in the civilisation of the ancient Rome and the ancient Greek. The stress on forms is the most important content of this style. The furniture design focuses on the sense of order, which prefers to arrange the furniture in line rather in scroll or curve. Furniture is decked with delicate paintings and carvings, or inset with gold foil. The best selection to present the neo-classic style includes chairs, tables and cabinets.

The chairs designed by British designer George Hepplewhit are the most typical for the neo-classic style presentation. The style of Hepplewhit's chairs is elegant, sleek and

家具是酒店中必备的器具，也是酒店设计中的重要内容，对室内的环境有很大影响。新古典风格家具来源于古罗马和古希腊时期的文明，对形式的强调是这种风格最重要的内容。家具注重秩序感，更多地使用直线，而很少使用卷轴和曲线。家具上带有精美的绘画和雕刻，另外，还会嵌入金箔等材料。最能表现出新古典风格特点的有椅类家具、桌类家具和柜类家具这三种。

新古典风格的椅类家具以英国设计师乔治·赫普怀特（George Hepplewhit）最为典型。赫普怀特式的椅子风格典雅优美、线条流畅，并且形式多样。尤其是椅背的设计非常丰富，有圆形、心形、椭圆形、盾牌形、环形、梯形等。其中，盾牌形的椅背最为著名，这种椅背中间镂空，并饰以各种图案，如麦穗、竖琴等，椅背底部有两根弯曲的木材连接座位的框架，扶手呈弯曲形与椅座连接，前腿为方形的直腿，由上到下逐渐变细，后腿为弯曲腿。这种盾牌形的靠背椅是赫普怀特风格家具中最为突出的作品，在美国流行一时。赫普怀特的椅子常用桃花心木、椴木、枫木做材料，技术精良，雕刻细致，很受当时人们的喜爱。

Elegant table
精美的桌子

Chapter Four
FURNITURE IN NEO-CLASSICAL HOTELS

diverse. In particular, the design for the back of chairs is a kaleidoscope of geometrical shapes like circle, oval, ring, trapezium, heart, shield, etc. Among them, the shield-shaped chairs are the most renowned, of which backs are pierced and decorated with various patterns, such as wheat, harps, etc. At the bottom of the back, two crooked wooden sticks connect the frame of seat; bent arms link to the seat; square forelegs are straight, and slim down from top to bottom; rear legs are bent. The shield-shaped back chair is the most outstanding work of the Hepplewhit collection, which was thriving in America for a period. Hepplewhit chairs are usually made by mahogany, basswood and maple with sophisticated technology and delicate carvings, which was favoured by people at that time.

Tables refer to dining tables, side tables, secretaries, tea tables, etc. The works of Tomas Sheraton, a British furniture designer, are the most typical. The design style of Sheraton collection is influenced by Louis XVI style and Adam style, emphasising lightness and simplicity. His works has a various catalog and strong practicality. They are always inset with drawers and caddies for

桌类家具有餐桌、墙边桌、工作台、茶几等。其中以英国家具设计师托马斯·谢拉顿（Thomas Sheraton）的作品最有代表性。谢拉顿家具的设计风格受到了路易十六风格和亚当风格的影响，强调轻便和简朴。他设计的桌类家具样式众多，并且实用性都很强。桌子中往往有暗藏的抽屉、小盒子等空间可以存储办公用品和日用品等，另外还有能拉出镜子的梳妆台。一种桌子还常常有多种用途，例如梳妆台可以兼作书架，书架带有书桌的功能等。除了实用性之外，谢拉顿的家具外形也很优雅，桌腿一般上粗下细，并带有复杂的雕刻，桌腿间常采用横档设计，台面用珍贵的木材，并有彩色的装饰，上面绘有人物、植物等图案。

赫普怀特的著作《橱柜制作师及软包师指南》（The Cabinet-Maker & Upholsterer's Guide）中展示了许多复杂的柜类设计，包括书橱、衣橱

Living room of California Cottage Suite
加利福尼亚别墅套房客厅

storing stationery and commodity. Besides, dressers installed a mirror that can be pulled out are common. One kind of table also usually has various functions, for example, dressers can be used as bookshelves, which can replace the desks. In addition to the functionality, the furniture by Sheraton is apparently elegant: the table legs slim off from top to bottom and carved with complicated patterns, between which crosspieces are connected; table tops is made by precious timber and decked with coloured figures and plants patterns. Hepplewhit's book – The Cabinet-Maker & Upholsterer's Guide, showcases a variety of cabinets, including bookshelves, wardrobes, etc. Generally, the deep drawers are installed in the wardrobes, and configured the spaces for toiletries. Such furniture is more delicate for catering to the women's tastes, so it is usually bordered with other materials and crooked surface.

In today's neo-classical hotels, such classical furniture is still popular. Every type of imitations of the neo-classical furniture add the classical atmosphere to hotel space in varied environment.

等。书橱的设计为了配合当时的贵族通常都很高大，分上下两层，上层带有玻璃花格门，下层根据不同的需求做成抽屉等不同的结构。为了强调装饰效果，表面还会带有一定的图案。衣橱类的家具一般带有很深的抽屉，并且带有专门摆放盥洗用具的空间。这类家具为了迎合女性的需求往往会装饰得更加精美，采用镶边、曲面等方式。在如今的新古典风格酒店中，这些经典的家具依然受到欢迎，各种类型的新古典家具仿制品在不同的环境中为酒店空间增加了古典的气韵。

Living room of Cottage Suite
别墅套房客厅

Chairs
椅类家具

Sitting on the Art Pieces
坐在艺术品之上

Chairs of the neo-classical style furniture have a large catalog, as well as rich forms and functions. Featured by their linear moulding, chairs are mainly characterised by reasonable structures and elegant decor. After centuries of development, the chair design, by the neo-classic period, had taken a thorough consideration of the most appropriate scales involving the height and the length of arms and the back was measured accurately. The selected materials for the chair back became more exquisite and softer. Chairs of the French Louis XVI period were a model. The chair legs adopted linear form with narrower bottom, and most of the chair backs are square. The arms are decked with cushion and connect the forelegs. The surfaces of such chairs employed superior silks, velvet or other materials, on which were embroidered plants, stripes and so on. Apart from arm chairs, sofas are more popular, especially the four-seat sofas similar to the armchairs. Chateau Mcely is a beautiful castle hotel. The interior design inherits the elegant and the classical European style, as well as the gorgcows furniture. In the bedrooms and restaurants, the neo-classical style armchairs can be seen everywhere, but they are simpler without complicated gilding. However, chair legs and arms are carved with simple patterns. The sofas, the same as arm chairs, match with the environment of bedroom perfectly. The design of chairs and sofas present the spirit of classicism and the hotel's characteristics of its own, making a deep impression to the guests.

椅类在新古典风格家具中种类很多，形式和功能也都很成熟，其主要特点是结构合理、装饰文雅、以直线造型为基础。椅类家具在经历了几个世纪的发展后，到新古典时期已经充分考虑了人体最舒适的尺度，在高度、扶手长度、椅背长度等方面都经过精确的测量。椅面的选择也更加考究，更加细软，尤其是法国路易十六时期的椅子堪称典范，椅腿采用直线形式，底部较细，椅背多为方形，扶手处带有软垫，与椅子前腿相连，椅面采用高档的绸缎、天鹅绒等材料，上面带有花草、条纹等图案。除了扶手椅之外，沙发也很普及，尤其是四人沙发很受欢迎，设计特点与扶手椅大致相同。

梅斯丽城堡酒店是一家漂亮的城堡酒店，酒店的内部设计秉承了欧洲风格的优雅和古典，家具也同样美轮美奂。在一些卧室和餐厅中随处可见新古典风格的扶手椅，这种椅子采用了和室内环境相同的颜色，并带有大朵的花纹，样式来源于法国的路易十六风格，但却更加简化，没有了复杂的镶金描银，但在椅腿和扶手等处也带有简单的雕刻。卧室中的沙发也与扶手椅如出一辙，与卧室的环境十分搭配。椅子和沙发既体现了古典的精神，也符合酒店自身的特色，足以让人印象深刻。

Chateau Mcely

梅斯丽城堡酒店

Location: Mcely, the Czech Republic
Designer: Mr. Otto Bláha, Mrs Inez Cusmano
Photographer: Chateau Mcely
Area: 2,300m²

项目地点：捷克，梅斯丽
设计师：奥托·巴拉哈；伊内兹·卡斯马诺
摄影师：梅斯丽城堡酒店
面积：2,300平方米

Chateau Mcely is a five-star eco chic chateau hotel, the first and only one of its kind in the Czech Republic. The quality of service, beauty of the interiors and impressive setting in the surrounding countryside were conceived for the most demanding of clients. Chateau Mcely can offer its guests accommodations in 23 original rooms and apartments, elegant chateau salons, the renowned Piano Nobile restaurant with its outdoor patio, the 17th century Alchemist Club and Wine Cellar, the Mcely Spa for healing and relaxation, a library in the tower of the chateau, a rooftop observatory, a well-maintained English park with swimming in a natural bio-lake with a white sand beach, a multifunctional sports ground, children's playground, and a herb garden.

There are a total of 23 rooms and suites for guests at the chateau, every room as unique as the chateau itself. Each of them has its own style, name and is furnished with the utmost care and emphasis on every detail. It is almost as if you were staying in 23 different hotels.

The ground floor is called the Floor of the World. Here you can find such rooms as the mysterious Orient, elegant Europa, the magical Southern Seas, and the beautiful and stunning Legend with its enchanting bathroom and balcony. The first floor follows the theme of time. Here most of the rooms are named after the months of the year. The Floor of Time ends in the Mark Twain Luxury Suite and the Rilke Double Room (named after the poet Rainer Maria Rilke, whose patron was Princess Thurn-Taxis), which pays tribute to one of the most significant visitors in the history of the chateau. All the rooms, however, have one thing in common – outside their windows lies the natural beauty of the English park and the St. George forest.

123

1. The library houses 18th-century scientific instruments.
2. Suite Africa on the ground floor has original style.
3. A corner of Suite Africa

1. 图书室内摆放着18世纪的科学仪器。
2. 一层的南非套房拥有原始风格。
3. 南非套房的一角。

梅斯丽城堡酒店是一家五星级时尚生态酒店，是捷克首家、也是唯一一家同类型酒店。酒店的服务质量、优美的室内装饰以及周边迷人的乡村美景足以让最挑剔的客人流连忘返。

梅斯丽城堡酒店拥有23间原创客房和公寓、优雅的城堡沙龙、著名的诺贝尔餐厅（配有露台）、源于17世纪的炼金术士俱乐部和酒窖、梅斯丽休闲治疗SPA、堡顶图书室、屋顶·望台以及维护良好的英式花园（配有天然生态湖游泳场、白沙滩、多功能运动场、儿童操场和香草园）。

酒店为客人提供了23件客房和套房，每个房间都与众不同，拥有独立的名称，并且经过了精心装饰。23间客房会令你感觉置身于23家不同的酒店。

酒店一层被称为"环游世界层"，房间被命名为神秘东方、优雅欧洲、魔力南海、美丽传奇（套房配有高级卫生间和阳台）等。

酒店二层以时间为主题，大多数房间都以12个月份命名。"时间层"的两端是马克·吐温豪华套房和里尔克双人间（以诗人莱纳·玛利亚·里尔克命名，他的赞助人是索伦–塔克西丝公主），这两间房的命名都向酒店历史上最著名的客人表达了敬意。

然而，所有客房都有一个共同点——它们的窗外都是英式花园和圣乔治森林的自然美景。

1. Spa reception is spacious and creatively-designed.
2. Three therapeutic suite Honey, Silk and Pearls at the Mcely spa offers a relaxing and inspiring environment for an exceptional spa experience.

1. 水疗馆的前台设计新颖，宽敞明亮。
2. 梅斯丽水疗中心为非凡的水疗体验提供了放松和灵动的环境。

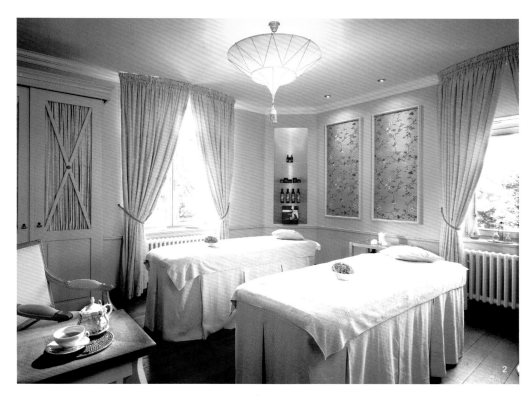

3. 传奇套房是酒店最奢华的套房，配有愉悦的浅色装饰；宾客一定会喜欢现代而华丽的19世纪风格浴室和俯瞰英式花园的小阳台。
4. 花朵房的主要特征在于墙面上手绘的应季花朵。
5. 如果将沙发作为沙发床，豪华套房可以充裕地容纳三个成人。

3. The Legend is the most luxurious suite in the chateau. It is harmonised with pleasant light pastel shades and guests will certainly be pleasantly surprised by the beautiful modern bathroom with 19th century styling and a small balcony with a view over the English park.
4. The dominant feature of Flower Room is a wall which is hand-painted with flowers typical for that month of the year.
5. The luxury suite can comfortably accommodate three adults by making use of the sofa which is also a day bed.

Tables
桌类家具

The World of Tables
桌上天地

Tables developed from the period of rack-foot style to the marble mosaic table of the Renaissance period, to the 17th century of British flap table, etc.. By the Rococo period, tables had developed tremendously. Besides common dining tables and desks, dressing tables, coffee tables, card tables appeared. When it came to the neo-classical period, on the basis of perfection, the table design was more refined and simpler in a more harmonious proportion. In hotels, the common tables include desks, dining tables, tea tables and so on. In the guestrooms, desks and tea tables have the similar forms with chairs. These tables use more straight lines, whose legs have more grooves and details are decorated simply. The whole design pursues appropriate proportion. The patterns of carvings have bouquets, leaves and fruits. Sometimes, there is a rail between the legs. The common materials are mahogany, ebony and basswood. However, in the restaurant, modern round tables are usually adopted. The Langham Huntington Hotel collects a variety of furniture, of which the neo-classical tables are the most distinctive. The desktops are all rectangular, and table bases are long and thin with carvings on the curved surface. The desk contains flexible drawers and bronze handles. Besides, there is nearly no more decoration. Simple as the structure is, it enjoys a harmonious proportion and its moulding is natural, perfectly matched with its surrounding furnishings, which highlight solemnness and magnificence. Compared with other furniture with complicated mouldings, the concise neo-classical style furniture is more suitable to modern hotel's requirements of comfort and high taste. Also, it is convenient to use.

桌类家具的发展经历了中世纪的架足式、文艺复兴时期的大理石镶嵌桌、17世纪英国的折叠桌等，到了洛可可时期，已经有了很大的发展，除了常见的餐桌、办公桌之外，还有梳妆桌、咖啡桌、牌桌等，到了新古典时期，酒店中，常见的桌类家具有办公桌、餐桌、茶几等。在客房中使用的办公桌、茶几一般与椅子的形式类似，这些桌类采用了更多的直线条，腿部一般带有凹槽，细部的装饰并不多，追求整体比例的适当，雕刻的图案有花束、叶形、水果，腿部有时带有横档，材料常用桃花心木、乌木、椴木。而在餐厅中，一般使用现代感较强的圆桌。

亨延顿朗廷酒店集合了许多类型的家具，其中，新古典风格的桌子最有特色。桌面都为直线矩形，桌脚细长且带有曲面的雕刻，办公桌带有灵活的抽屉和青铜把手，除此之外并没有过多的装饰，虽然结构简单，但桌子比例和谐，造型流畅自然，与周围的家具搭配协调，凸显出了庄重和华丽。与其他造型复杂的家具相比，这种简洁的新古典风格更适应现代酒店对舒适度和高品位的要求，使用起来也更加方便。

The Langham Huntington

亨廷顿朗廷酒店

Location: Pasadena, USA
Designer: The Johnson Studio
Photographer: The Langham Huntington
Area: 5,400m²

项目地点：美国，帕萨迪纳
设计师：约翰逊工作室
摄影师：亨廷顿朗廷酒店
面积：5,400平方米

The Langham Huntington, Pasadena is an iconic landmark hotel since 1907, located at the base of the picturesque San Gabriel Mountains, just minutes from downtown Los Angeles in beautiful Pasadena.

Originally the main entrance to the hotel, The Lobby Lounge offers panoramic views of The Horseshoe Garden and San Marino.

The Lobby Lounge décor features a mixture of modern furnishings and traditional finishes which present a uniquely warm and inviting space for guests to relax in luxury.

Huntington Suite: This delightfully appointed accommodation offers views of the gardens from the bedroom and a classic sense of style. The separate parlor area features comfortable seating, work desk and double French doors. The heritage opulence is complete with a graceful, four-poster bed and magnificent bathroom.

Langham Club Suite: Breathtakingly spacious, these suites are located on an exclusive key-accessed floor and include admittance to the private Club Lounge with a dedicated Concierge and five distinct culinary presentations daily. Separated from the bedroom by elegant French doors, the parlor area offers sumptuous seating, inspiring a feeling of utter contentment.

Cottage Suite: These generously sized rooms, situated in cottages throughout the hotel grounds, include a magnificently furnished bedroom and separate parlor featuring comfortable seating, work desk and an armoire. The atmosphere of timeless elegance ensures your every homecoming is a delight.

California Cottage Suite: Situated in cottages across the hotel grounds and in Royce Manor, some of these suites feature a fireplace, others a patio. All include a bedroom and separate parlor area featuring comfortable seating, work desk, an armoire and pleasing views of the grounds.

Clara Vista Suite: Located in the 1930s Spanish mission-style cottage Clara Vista, these lavish suites fascinate with hand-carved furnishings and a library that includes antique leatherbound volumes. The living room in each suite underscores the picture of domestic bliss with fine leather chairs, plush couches and a fireplace.

1. 酒店的中央楼梯平台配有可供人休息的沙发。
2. 宏伟的酒店大堂中央是巨大而华丽的水晶花瓶。
3. 亨廷顿图书室是酒店最吸引人的地方之一，里面藏有18、19世纪的艺术品和书籍。
4. 酒店新开的欧式风格大堂酒廊。
5. 大堂酒廊曾是酒店的大堂，酒廊的落地窗可以俯瞰U形花园和圣马力诺的美。

1. Centre landing with staircase has sofas for rest.
2. Large crystal vases of the luxurious flower centre-pieced the palatial lobby
3. The Huntington Library is one of the most popular attractions, has a fabulous collection of 18th and 19th century art and much more.
4. The hotel's recently opened European-style lobby lounge
5. As the hotel's original lobby, The Lobby Lounge features floor to ceiling windows with panoramic views of The Horseshoe Garden and San Marino.

Floor plan / 平面图

1. Terrace Restaurant / 1. 露台餐厅
2. Viennese Terrace / 2. 维也纳露台
3. Foyer / 3. 门厅
4. Viennese ballroom / 4. 维也纳宴会厅
5. Courtyard / 5. 庭院
6. The Royce / 6. 罗伊斯厅
7. Lobby lounge / 7. 大堂酒廊
8. The tap room / 8. 酒吧间
9. Georgian Ballroom / 9. 乔治亚宴会厅
10. Front desk / 10. 前台
11. Hotel lobby / 11. 酒店大堂
12. Gift shop / 12. 礼品店
13. Huntington foyer / 13. 亨廷顿门厅
14. Huntington ballroom / 14. 亨廷顿宴会厅
15. Colonnade / 15. 柱廊
16. Plaza / 16. 广场
17. Pavilion / 17. 休息亭
18. Entrance / 18. 入口

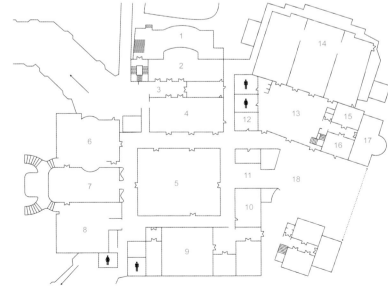

1. In the living room of Tourament of Roses Suite guest will discover a baby grand piano.
2. Located adjacent to The Viennese Ballroom and Terrace, The Boardroom is a luxurious and distinguished room situated for small meetings or speaker presentations.
3. With stained glass windows, gold-guild vaulted ceilings and Flemish art, The Georgian Ballroom offers more than a century of historic charm and provides an elegant setting for both meetings and meals.
4. The Huntington Spa at The Langham Huntington is now open following an exciting renovation and refurbishment. The new spa has a couple's room.
5. The corridor has great carved wall.

1. 玫瑰竞赛套房的客厅内摆放着一架小型三角钢琴。
2. 会议厅紧邻维也纳宴会厅，其风格奢华独特，适合举办小型会议或演讲。
3. 乔治亚宴会厅装饰着彩色玻璃窗、镀金拱顶和法兰德斯艺术品，以其百年的历史魅力打造了优雅的会议和就餐空间。
4. 亨廷顿水疗中心经过翻修重新开业。新建的水疗中心设有情侣房。
5. 走廊墙面上有着巨大的雕刻。

位于帕萨迪纳市的亨廷顿朗廷酒店建于1907年，是当地的地标性建筑。酒店坐落在圣盖博山脉的山麓，距离市中心只有数分钟的路程。

大堂酒吧曾是酒店主入口，从酒吧可以感受到U形花园和圣马力诺的美景。大堂酒吧内的现代家具和传统装饰交相辉映，呈现了独特的温馨氛围，让宾客们在奢华中体验休闲。

亨廷顿套房：亨廷顿套房采用古典风格进行装饰，享有花园的美景。独立的会客区以舒适的座椅、办公桌和法式双开门为特色。优雅的四柱床和华丽的浴室彰显了套房的经典和奢华。

朗廷俱乐部套房：朗廷俱乐部套房异常宽敞，集中安排在一个必须用钥匙进入的专属楼层。私人俱乐部酒吧配有专属门房，每天呈现五种独特的菜肴。会客区与卧室以优雅的法式门隔开，华丽的座椅让人心满意足。

村舍套房：这些宽敞的套房坐落在酒店领地的乡村小屋里，拥有装饰华丽的卧室和独立会客区（配有舒适的座椅、办公桌和大衣橱）。永恒的优雅氛围让人流连忘返。

加州村舍套房：套房位于酒店的乡村小屋和罗伊斯庄园，一些套房配有壁炉，另一些则有露台。所有套房都配有卧室和独立办公区，以舒适的座椅、办公桌、大衣橱和庭院的美景为特色。

克拉拉景观套房：这些豪华的套房位于19世纪30年代风格的克拉拉景观小屋内，配有手工雕刻的家具。套房的图书馆内藏有古董皮面书卷。套房的客厅具有家居氛围，配有精致皮椅、长沙发和壁炉。

1. Tournament of Roses Suite is loated on the Langham Club floor.
2. A separate dining room in Tournament of Roses Suite
3. Cottage Suite situated in cottages throughout the hotel grounds, include a magnificently furnished bedroom and separate parlor featuring comfortable seating, work desk and an armoire.
4. Spacious and sophisticated, Deluxe Room project the aura of a bygone era with tasteful period furnishings. Framed by ornate drapes, the windows allow the room to fill with natural light.
5. Separated from the bedroom by elegant French doors, the parlor area offers sumptuous seating, inspiring a feeling of utter contentment.

1. 玫瑰竞赛套房位于朗廷俱乐部楼层。
2. 玫瑰竞赛套房的独立餐厅。
3. 村舍套房坐落在酒店领地的乡村小屋里，拥有装饰华丽的卧室和独立会客区（配有舒适的座椅、办公桌和大衣橱）。
4. 豪华客房宽敞精致，以品位非凡的复古家具重现了过去的时代氛围；装饰着华丽窗帘的窗户让室内充满了阳光。
5. 客厅通过法式双开门与卧室隔开，提供了豪华的座椅，让人非常舒适。

Cabinets
柜类家具

The Prosperous Corner
角落里的繁华

In hotels, closets are the most common cabinets. Neo-classical closets have various sizes and mouldings. Generally, with two to three big drawers, the outlines are rectangular or semicircular and its corners are right angles. The handles of closets are decorated with gilding or bronze pendants, and the surface is carved with flora patterns. Sometimes, the goddess figures decorate the corners. The legs of closets are generally short, most of which is reverse cone-shape. Besides closets, the locker with high legs is common as well. The tallboy which can be removed has more drawers and is convenient for use. The decorations are similar to the closets', also with bronze handles and inlayed veneers. But the legs are taller and generally curved.

In Grand hotel Vesuvio, there is a large number of cabinets. In one of the suites, the lockers are very gorgeous. The whole locker is composed of two parts. The lower rectangular cabinet contains four drawers, of which door is inlayed with oil paintings. Through the glass door of upper locker, the delicate porcelain can be seen. The connection between the glass doors is sculpted into Ionic order, looking pristine and solemn. In the middle of its upper part of the whole locker, the cabinet has eagle-shaped ornament full of empire style. Besides the grandeur and gorgeous locker, there are many smaller cabinets in the hotel. Without overmuch ornament, this kind of cabinet inherits the concise style of the neo-classical style. Only on the surface, there are simple carvings and metal pendants. However, the overall proportion is appropriate and convenient for use. The cabinet can be located alongside the head of beds or the corner of walls.

在酒店中，衣橱是最常见的柜类家具。新古典风格的衣橱有各种不同的大小和造型。轮廓通常为矩形或半圆形，转角处呈直角，带有2~3个大抽屉，把手处通常是由镀金青铜制成的挂件，柜子表面带有花朵、垂饰等图案，有时在转角处还带有女像柱装饰。衣橱的腿一般很短，多呈上大下小的锥形。除了衣柜之外，一种腿部较高的储物柜也很常见。这种高脚柜抽屉较多，并且可以拿起来，使用很方便，装饰与衣橱类似，也带有青铜把手、镶嵌贴面等。但腿部较高，并且一般做成弯曲形。

维苏威大酒店中的柜类家具有很多，其中一间套房中的储物柜非常华丽，整个储物柜分成上下两个部分，下面的矩形柜带有四个抽屉，门上有镶嵌的油画装饰。上面的储物柜带有玻璃门，可以看到里面摆放的精美瓷器，玻璃门中间连接处被雕刻成了爱奥尼克柱式的形式，显得古朴庄重。整个储物柜上方中间部分带有鹰形的装饰，有着帝国风格的色彩。除了这种高大华美的储物柜之外，酒店中还有很多小型的柜子。这种储物柜沿袭了新古典风格的简洁，没有过多的装饰，只是表面带有简单的雕刻和金属挂件，但整体比例适当，使用起来也很方便，适合在酒店中摆放在床头或墙角的位置。

Grand Hotel Vesuvio

维苏威大酒店

Location: Naples, Italy
Architects/Designers: Architect Sergio Bizzarro
Photographer: Grand Hotel Vesuvio
Area: 16,800 m²

项目地点：意大利，那不勒斯
设计师：塞尔吉奥·比扎罗
摄影师：维苏威大酒店
面积：16,800平方米

Right on the sea front, opposite the picturesque Santa Lucia Harbor, Grand Hotel Vesuvio Naples is one of the Italy's finest luxury hotels. The hotel first opened its doors to a selected and discerning international clientele in 1882 and for many decades represented the most sought after venue in the city where to host glamorous social gatherings and important cultural events.

The second world war resulted in serious damage to the city of Naples and the Grand Hotel Vesuvio Naples was just one of the buildings to be almost completely destroyed in heavy bombing raids. In the immediate post war years, the hotel was rebuilt and two additional floors were added before the Grand Hotel Vesuvius' grand reopening in 1950.

The Grand Hotel Vesuvio Naples has 163 bedrooms and 17 suites, all beautifully furnished and equipped with every modern comfort. The majority of the rooms have charming balconies or terraces from where to admire spectacular views of the Bay of Naples, Mount Vesuvius, and the island of Capri.

One of Naples' most highly acclaimed restaurants, the "Caruso Roof Garden", named after the famous tenor who regular sojourned at the Grand Hotel Vesuvio Naples, provides a supremely elegant dining venue where to savor excellent Mediterranean cuisine and any of an extensive selection of prestigious wines. Located on the 9th floor of the Grand Hotel Vesuvio Naples, the restaurant opens out on to a panoramic terrace which, during the long warm Neapolitan summer, transforms in enchanting al fresco dining room. Grand Hotel Vesuvio Naples is renowned for its excellent organisation of high profile conferences and meetings; offering a state-of-the-art business center and 7 well equipped function rooms for as many as 400 people.

坐落在海滨的维苏威大酒店正对风景如画的桑塔露琪亚海港，是意大利最精致的奢华酒店之一。1882年开业时起，维苏威大酒店在那不勒斯城中长盛不衰，在数十年间举办了众多的社交集会和文化活动。

第二次世界大战对那不勒斯造成了严重的破坏，维苏威大酒店也在空袭轰炸中几乎被摧毁。战后，酒店进行了重建并新增了两层楼。酒店于1950年重新盛大开业。

那不勒斯维苏威大酒店的163间客房和17间套房全部采用了精美的装饰，配有现代化服务设施。大多数房间都设有迷人的阳台或露台，为宾客提供了那不勒斯湾、维苏威火山和卡普里岛的美景。

作为那不勒斯最负盛名的餐厅之一，卡普索屋顶花园餐厅以曾居住在维苏威大酒店的著名男高音的名字命名，为宾客们提供了一个享用地中海美食和美酒的优雅就餐场所。餐厅位于酒店的9层，在那不勒斯的漫漫长夏，宾客可以在宽阔的露台上享用美食。

维苏威大酒店以其优秀的会议和活动组织能力而著称，酒店拥有一个配有先进设施的商务中心和7间可容纳400人的功能厅。

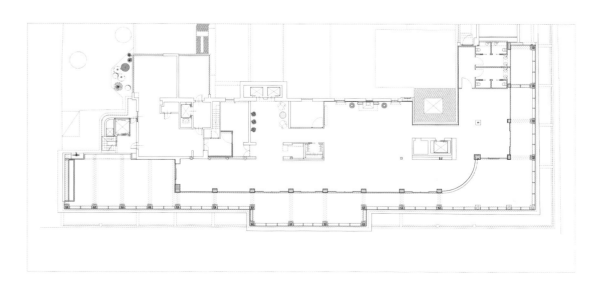

Floor plan
1. WC
2. Corridor
3. Staircase
4. Terrace

平面图
1. 卫生间
2. 走廊
3. 楼梯
4. 露台

1. Sitting room of Caracciolo TC Suite
2. Deluxe Sea View rooms feature 19th century furnishings in a local style with a modern touch.
3. Junior Suite is tastefully furnished with original period furniture.
4. Deluxe Seaview room has a seating area with sofa and a balcony.

1. 卡拉乔套房的客厅。
2. 豪华海景客房的19世纪风格的装饰结合了当地风格和现代装饰。
3. 普通套房装饰着古典家具。
4. 豪华海景套房的会客区配有沙发和阳台。

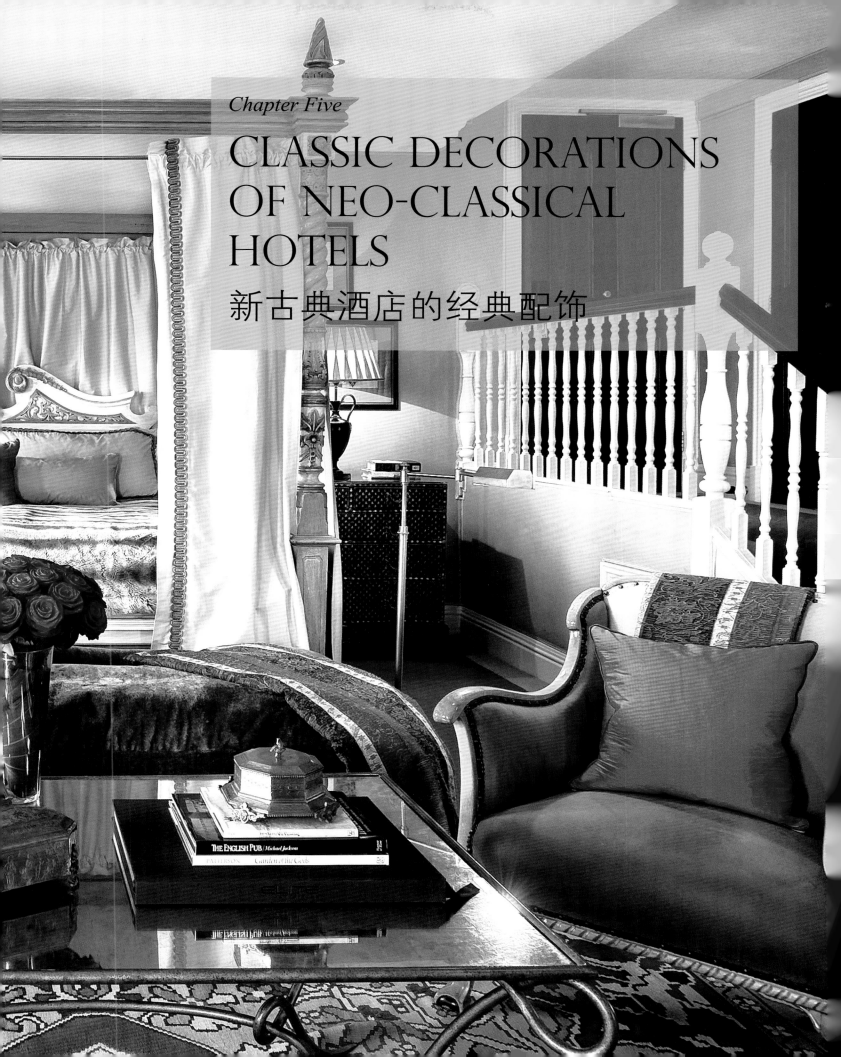

Chapter Five

CLASSIC DECORATIONS OF NEO-CLASSICAL HOTELS
新古典酒店的经典配饰

Chapter Five
CLASSIC DECORATIONS OF NEO-CLASSICAL HOTELS
新古典酒店的经典配饰

Some elements for creating the neo-classical style are necessary, such as classical orders, fireplaces, luxurious textiles and crystal lamps, all of which are essential in the neo-classical style.

The classical order refers to the order in the period of ancient Greece and ancient Rome. The orders in ancient Greece include the Doric order, the Ionic order and the Corinthian order. These three kinds of orders were inherited in ancient Rome, and then developed into the Tuscan order and the Composite order. At first, these classic orders are applied in the architecture, and then gradually are used in the interior decor, even in the furniture design. The neo-classical style usually adopts these three most common orders in the period of ancient Rome and ancient Greece, especially the Ionic order and the Corinthian order.

The hotel lobby is the area that applies orders frequently. Magnificent and grandeur marble columns can set off the solemn atmosphere by contrast, which is suitable for the lobby with large area. The comparatively small lobby can use attached columns that are buried partly in the wall, or use plat pilasters.

The fireplace originates from the Oc-

新古典风格的营造离不开一些固有的元素，古典柱式、壁炉、华丽的织物、水晶灯，这些都是新古典风格中几乎不可缺少的东西。

古典柱式是指古希腊和古罗马时期的柱式。古希腊时期的柱式有陶立克柱式、爱奥尼克柱式、科林斯柱式。古罗马时期继承了这三种柱式，另外还发展出了塔司干柱式和混合柱式。这些经典柱式最初被应用在建筑中，后来被渐渐用在室内装饰中，甚至于家具设计中。新古典风格中常用到的是古希腊和古罗马所共有的三种柱式，尤其是爱奥尼克柱式和科林斯柱式。

酒店大堂是柱式应用最频繁的区域，雄伟高大的大理石柱能烘托庄重的气氛，适合面积较大的大堂。而相对较小的大堂可以使用将柱子部分埋入墙中的半柱，或把柱子做成扁平状的壁柱。

壁炉起源于西方，可以称得上是欧洲的文化符

The lobby's beautiful marble floor and sweeping double staircase weclome guests to finest rooms.

大堂美丽的大理石地板和两个大面积的楼梯欢迎着客人走进客房。

Chapter Five

CLASSIC DECORATIONS OF NEO-CLASSICAL HOTELS

cident, which can be called the symbol of the European culture for its self-evident importance in the interior design. It can be seen everywhere in guestrooms, restaurants, ballrooms, etc. The forms of fireplaces are various, such as marble fireplaces, wooden fireplaces, Parian marble fireplaces, etc. for different interior environment. As for the neo-classical style, the fireplace design is generally concise. Most of the fireplaces are made by marble and centred in guestrooms.

Luminary is finishing touch in the interior design. Crystal lamps are gorgeous and various, considered as the best choice in the neo-classical style. Crystal lamp originated from 17th century when the rococo style was prevailing and luxurious crystal lamps were popular. Nowadays, crystal lamps have more and more mouldings, and are more and more intricate. The chandelier in the ballroom of the hotel is usually composed by thousands of components. Besides crystal, luminaries use gold, silver, copper and other metals, which plays the role of connections,

号，在室内装饰中的作用不言而喻。在酒店的客房、餐厅、宴会厅等几乎随处可见。壁炉的形式有很多种，大理石壁炉、木制壁炉、仿大理石壁炉等，可以适合不同的室内环境。在新古典风格中，壁炉的设计一般比较简洁，以大理石壁炉为主，通常都设在房间的中心位置。灯具的搭配通常是室内设计中的点睛之笔，水晶灯雍容华贵并且种类繁多，是新古典风格中不二的选择。水晶灯起源于17世纪的欧洲，也就是洛可可风格风行的时代，华丽耀眼的水晶灯受到很多人的欢迎。如今，水晶灯的造型越来越多，也越来越复杂，酒店大型宴会厅中所用的大型灯一般都由几千个构件组成。除了水晶之外，灯具中还会带有金、银、铜等金属，起到连接的作用之

Vintage Cocktail Bar featuring warm tones of mahogany lined interior and rich fixtures.

老式鸡尾酒吧以暖色调的桃花心木内饰和丰富的家具为特色。

and at the same time adds the sense of luxury. The neo-classical style crystal lamps highlight the beauty of symmetry and reasonable proportion. They can be used in most areas of the hotel except that in some places where strong light is required. Textiles includes bed linens, curtains, carpets, slip covers, cushion, etc. The masterly application of textiles can enrich the dull interior design and make the design full of interests. The texture, colour and pattern of textiles provide unique experience, presenting a certain degree of cultural connotation. The neo-classical textile style features elegance and gorgeousness. Slip covers, curtains, cushions and other texiles. adopt velvet, silk, satins and other materials with soft and smooth textures. The bed linens requir more comfort adopt chintzes or soft silks. The colours are mainly gold, yellow, olive-green, crimson, dark brown, mazarine blue, etc. And carpets and tapestries are often woven by wools.

外，更增添了华丽之感。新古典风格的水晶灯强调对称的美感，以及合理的比例，在酒店的大部分地方都可以使用，只是在一些对光线要求较高的地方需要搭配其他光源使用。

织物包括各种床上用品、窗帘、地毯、沙发套、靠垫等。织物的巧妙运用可以使静态单调的室内空间变得丰富多彩充满情趣，织物的质感、颜色、图案也能让人感受到不同的体验，体现出一定的文化内涵。新古典风格的织物风格以典雅、华丽为主要特点，沙发套、窗帘、靠垫等应用天鹅绒、绸缎等质感柔软、光洁度强的织物较多，对舒适度要求较高的床上用品更多采用印花棉布或柔软的丝绸，色彩主要有金色、黄色、橄榄绿、深红色、深棕色、深蓝色等。而地毯挂毯一般由羊毛织成。

The lobby has high ceiling and cozy chairs.

大堂有高高的天花板和舒适的椅子。

Order
柱式

Various Forms of Oders
当柱式的形式有更多可能

The order which can adapt to different environments has rigorous, delicate and various forms but made in a simple way and is applied widely. Besides the five basic forms, the order forms can be changed appropriately according to practical situations in hotels, such as increasing the ornaments on the chapter, changing the strix, or adding more colours to the fust. The texture of the orders can produce ornamental effect to a certain extent. Vigorous and luxurious marble orders are applied widely; softer and warmer wooden orders are usually set in small restaurant. Besides a single order, pilasters hidden in the wall and attached columns can also decorate and divide the space, saving part of the space as well.

In the lobby of Fairmont Grand Hotel Kyiv, the marble columns around the walls are extraordinarily eye-catching. These orders are not traditional classical order. The chapiter patterns on the Ionic order and the underpart is decorated with gilded pendants, echoing the surrounding carved gilding. Adopting colourful marbles, the smooth fust with strixes is more distinctive on the white and yellow walls. These orders enhance the neo-classical style of the lobby and integrate with the surrounding environment skillfully. In the restaurant of Fairmont Grand hotel Kyiv, walls and ceilings are made of wood and accordingly, such materials are also employed by the order. Its chapiters is carved with scrolls of the Ionic order, and the fust has strixes but does not spread to the top. The whole column is short and simple. However, it exerts a strong influence on the decorative effect in the narrow and small space.

柱式这种结构形式严谨、精致、技术简单，形式多变，能适应不同的环境，因此应用也很广泛。在酒店中，除了古典柱式的五种基本形式之外，还会根据实际情况对柱式做出相应的改变，增加柱头上的装饰、改变柱身凹槽，或者在柱身表面增加色彩等。柱式本身的材质也会产生一定的装饰性，大理石柱式雄浑华丽，应用最为广泛，木制的柱式更加柔和温暖，适合一些小型的餐厅。除了单独使用之外，隐藏在墙面中的壁柱和半柱形式也可以起到装饰和分割空间的作用，同时也能节省一部分空间。

基辅费尔蒙特大酒店的大堂区中，周围紧靠墙壁的大理石柱异常夺目。这些柱式并不是传统的古典柱式，柱头以爱奥尼克柱式为原型，但在下面增加了镀金的挂件装饰，与周围的镀金雕刻交相辉映，柱身光滑没有凹槽，采用了彩色大理石为材料，在周围以白色和金色为主的墙面中显得更加突出。这些柱式有效地强化了大堂的新古典风格，与周围环境的结合也很巧妙。而在基辅费尔蒙特大酒店的餐厅中，墙面和天棚都用了木制的材料，柱式也相应地使用了相同的材料。柱头是爱奥尼克柱式形式的卷轴，柱身带有凹槽，但没有一直延伸到顶部。整个柱子短小简洁，却在狭小的空间中起到了很强的装饰作用。

Fairmont Grand Hotel Kyiv

基辅费尔蒙特大酒店

Location: Kyiv, Ukraine
Designer: Duangrit Bunnag
Photographer: Fairmont Hotels&Resorts
Area: 15,000m^2

项目地点：乌克兰，基辅
设计师：都安格力特·巴纳格
摄影师：费尔蒙特酒店度假村集团
面积：15,000平方米

Opened in March 2012, the classically designed Fairmont Grand Hotel Kyiv features 258 rooms, including 54 suites offering an elegant and relaxing atmosphere as well as Royal and Presidential suites.

The Atrium Lounge & Pâtisserie, with its authentic art nouveau design, offers the guests the perfect place to enjoy a traditional afternoon tea or choose from a wide selection of coffees and loose leaf teas.

Located in the Hotel Lobby, Grand Cru Champagne Bar features an impressive selection of Champagnes by glass or bottle. The elegant proportions and stately procession of rooms complement the sumptuous furnishings and beautiful artwork to create a Royal Suite that is literally fit for a king. Guests enter through a marble foyer, off of which opens a beautiful wood-panelled office with inspiring views of Kyiv and stylish office equipment. Guests then proceed through to a comfortable sitting room and a separate dining room that seats up to eight guests in a light, airy interior influenced by Ukrainian formal gardens. The master bedroom is positioned adjacent to a large, walk-in dressing room with cedar-lined wardrobes & a spacious, ensuite, his & hers bathroom with a deep, soaking tub and a steam shower. These one-bedroom suites offer stunning, panoramic Dnipro River views. Decorated in elegant style, they feature marble entrance foyer, plush sofa and two armchairs in the sitting room, and an armchair in the bedroom, personal bar and flat-screen televisions. The well-appointed marble bathrooms feature separate walk-in showers with rain showerhead, large chrome fittings and Le Labo amenities. Fairmont Gold rooms overlook the quiet Podil area, the inner courtyard or Dnipr River. The well-appointed marble bathrooms all feature separate walk-in showers, large chrome fittings and Le Labo amenities

基辅费尔蒙特大酒店于2012年3月正式开放，酒店采用古典设计，拥有258个房间，其中包括54间环境优雅、舒适的普通套房和皇室及总统套房。

中庭休息区和法式甜点屋以其卓越的新艺术风格设计为客人提供了享用传统下午茶的完美场所，同时还提供多种多样的咖啡、茶叶供客人选购。

酒店大堂的顶级酒庄香槟吧为宾客提供了各色香槟，宾客可以单杯购买，也可以成瓶享用。

优雅的比例，排列整齐的房间，辅以华丽的家具装饰和美丽的艺术品，皇室套房将为宾客提供国王般的体验。穿过大理石门厅，映入眼帘的是以木质镶板装饰的办公室。办公室配有时尚的办公设备，从窗户可以看到基辅的美景。往里走是舒适的客厅和独立的餐厅。餐厅可容纳8人，室内环境清新明亮，深受乌克兰花园设计的影响。主卧与一间大型步入式化妆间（配有松木衣柜）和宽敞的浴室（配有舒适的浴缸和蒸汽淋浴）相邻。

单卧套房则享有第聂伯河的美景。套房优雅的风格以大理石门厅、长毛绒沙发为特色。客厅内设有两张扶手椅，卧室内也设有一张。套房同时还配有私人吧台和平板电视。设施完善的大理石浴室内有独立的步入式淋浴间，配有淋浴头、铬合金配件和勒拉伯卫浴设施。

费尔蒙特黄金客房俯瞰着宁静的波蒂区、内庭或是第聂伯河。设施完善的大理石浴室全部配有独立的淋浴间、铬合金配件和勒拉伯卫浴设施。

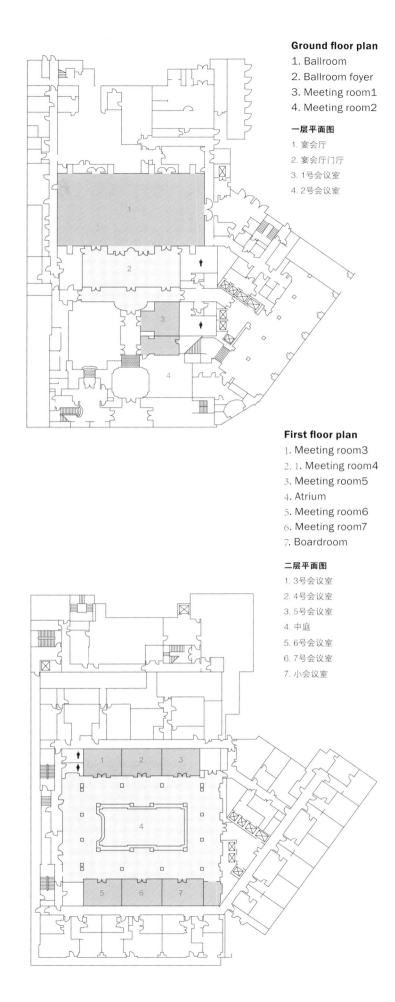

Ground floor plan
1. Ballroom
2. Ballroom foyer
3. Meeting room1
4. Meeting room2

一层平面图
1. 宴会厅
2. 宴会厅门厅
3. 1号会议室
4. 2号会议室

First floor plan
1. Meeting room3
2. Meeting room4
3. Meeting room5
4. Atrium
5. Meeting room6
6. Meeting room7
7. Boardroom

二层平面图
1. 3号会议室
2. 4号会议室
3. 5号会议室
4. 中庭
5. 6号会议室
6. 7号会议室
7. 小会议室

1. The new lobby designed beauty and complete with a sitting area
2. The reception of meeting room is elegance and charm.
3. Located in the hotel lobby, Grand Cru Champagne Bar features an impressive painting.

1. 新大堂设计美观，设有休息区。
2. 会议室的前台优雅而迷人。
3. 位于酒店大堂的顶级酒庄香槟吧以精美的油画为特色。

1. Occupying an entire wing on the eighth floor, the Presidential Suite is among the Ukraine's most prestigious accommodations, it has a very spacious sitting room.
2. All Junior Suites offer a perfect combination of classic décor and contemporary amenities. Panoramic views of the city are afforded through large windows.
3. Bedroom of the Presidential Suite has Empire Style.
4. Dnipro Signature Suite is a spacious comfortable room that consist of separate bedroom with King sized Stearns and Foster bed and well appointed sitting room and Balcony.

1. 总统套房占据了9层的一侧，是乌克兰最负盛名的酒店客房之一；它拥有宽敞的客厅。
2. 所有普通套房的设计都完美融合了古典装饰和现代设施；大窗户让城市全景尽收眼底。
3. 总统套房的卧室采用了帝国风格设计。
4. 迪尼普极品套房宽敞舒适，配有独立卧室（内置特大的四柱床）、设施齐全的客厅和阳台。

Fireplace
壁炉

When the Fireplace Becomes the Finishing Touch in the Interior
当壁炉成为室内的点睛之笔

Fireplace, an important element in the European style, is appropriate for any type of guestroom. Its decorations are always the centre of the interior design and have tremendous influence on the interior design style. The neo-classical fireplace is decorated delicately. Whether on the marble or wooden fireplace, the mantels are all carved with patterns, of which the most common ones are floral motives and order motives. In Europe, real flame fireplaces are usually adopted, as it has both functional and decorative features. In hotels, the fireplace plays a decorative role, so its design usually focuses more on the match with the interior design, especially the wall design. White marble fireplace matching with panel walls is the most classic combination in the neo-classical style, especially in the guestroom of hotel. In addition, the decoration above fireplace is also important, and oil painting is a frequently-used choice.

The guestrooms and the library in Bovey Castle are all decked with fireplaces. Fireplaces in guestrooms adopt white marble mantel carved with scrolled patterns. Above the mantel are porcelains and other artworks. The pediments above the fireplace are inlayed square mirrors, around which are floral patterns. The whole mantel and pediment are white in accordance with the white wainscot in the room, and the walls are painted blue, making the whole room simple and elegant. In the library, the fireplace adopts wooden mantel carved with floral patterns, and pediments with geometry patterns, echoing the plaster ceiling. The wood-coloured fireplace is not only solemn and elegant, but creates a warm atmosphere.

壁炉是欧式风格中的重要元素，适合各种类型的房间，并且壁炉的装饰一般都是室内的核心，对室内的装饰风格有很大影响。新古典风格的壁炉装饰精美，无论是大理石壁炉还是木制壁炉，壁炉架上都会有雕刻的花纹，常见的雕刻有花朵图案或是柱式形式。在欧洲地区，一般会采用真火壁炉，兼具使用功能和装饰功能，在酒店中，壁炉主要起装饰的作用，因此更注重与室内风格尤其是墙面的搭配。白色的大理石壁炉搭配镶板的墙面是新古典风格中经典的搭配，尤其适合在酒店的卧室中使用。另外，壁炉上方的装饰也很重要，油画是常用的选择。

波维城堡酒店中的客房、图书馆都带有壁炉。客房中的壁炉采用了白色大理石的壁炉架，带有卷轴形的雕刻。壁炉架上方摆放着瓷器等艺术品。壁炉上面的山花上镶嵌着方形镜子，周围是花形的雕刻。整个壁炉架和山花全部是白色，与房间内白色的墙裙统一，而墙面也装饰成了淡蓝色，整个房间整洁而素雅。图书馆中的壁炉使用了木制的壁炉架，壁炉架上采用了复杂的花朵雕刻，山花上还有几何形状的雕刻，与石膏天花板相互呼应。木色的壁炉不仅庄重典雅，也创造了温馨的氛围。

Bovey Castle

波维城堡酒店

Location: Exeter, UK
Designer: Annabel Elliot
Photographer: Bovey Castle
Area: 5,000m²

项目地点：英国，埃克塞特
设计师：安娜贝尔·艾略特
摄影师：波维城堡酒店
面积：5,000平方米

Bovey Castle is situated in the beatiful Dartmoor National Park, in the heart of Devon. Stylish, contemporary and unashamedly indulgent, the Bovey Castle estate hides 14 three-storey country lodges within its grounds, all in keeping with the environment beyond. Each lodge has three en-suite bedrooms and almost all are laid out over three floors, sleeping up to eight people in total. Each lodge features three, double en-suite bedrooms, a guest bathroom, open-plan kitchen, lounge, dining area and a fully-fitted oak kitchen with Belfast sink.

Fashioned from local granite and vaulted with English oak, each property echoes the enduring character of Devon's ancient moorlands. Shaded by woodland, bordered by lakes and manicured lawns, the lodge becomes your own private sanctuary – set within more then 400 acres of Devon countryside and minutes away from the luxurious hospitality of the five-star Bovey Castle.

Bovey Castle features 64 individually designed bedrooms by Annabel Elliot, located in the original manor house and private mews. Each bedroom reflects the luxurious hospitality that Bovey Castle is renowned for, with every finishing touch selected to complement the warmth and elegance of our guest rooms.

The dining experiences at Bovey Castle have been designed to complement the hotel's reputation for country house elegance and outdoor activities. Whether you plan to participate in one or two of the hotel's impressive guest activities or retreat to the tranquility of the Adam Room's plush armchairs, the style and selection found in each of the hotel's restaurants and dining areas cater for all tastes.

161

波维城堡酒店位于德文郡的中心——风景优美的达特穆尔国家公园。

时尚、现代而又奢华的波维城堡酒店度假区内拥有14座三层高的乡村别墅，它们与周边环境紧密相连。每座别墅都有三间套房，分布在三个楼层，最多可容纳8人。每座别墅有三间双人套房、一个客用洗手间、开放式厨房、休息室、就餐区和设施齐全、配有水槽的橡木厨房。

由当地大理石和英国橡木建成的别墅反映了德文郡古代沼泽高地坚忍的特征。在绿树成荫，碧水青草的环绕之中，别墅将成为你的私人度假胜地。别墅坐落在占地400多英亩（约162公顷）的德文郡乡村，距离奢华的五星级波维城堡酒店仅有几分钟的路程。

波维城堡酒店拥有64间独立设计的客房，全部由安娜贝尔·艾略特操刀完成。酒店位于一座古老的庄园内，配有私人马厩。每间客房都彰显了波维城堡的奢华品质，每件装饰都致力于打造温馨、优雅的客房体验。

波维城堡酒店的餐厅体现了酒店的乡村优雅，彰显了其户外特色。无论是参与酒店出色的游客活动之中，还是在亚当餐厅柔软的座椅上静享片刻安宁，酒店餐厅和各个就餐区的精美风格和特色美食能够满足每位客人的需求。

Hotel plan

1. Croquet Lawn
2. Activities Area
3. Summer House
4. Long Barn
5. Tennis Courts
6. Activities Office
7. Playroom
8. The Mes Conference Centre
9. Petanque Court
10. Bistro Terrace
11. Swimming Pool
12. Golf Shop

酒店平面图

1. 门球草坪
2. 活动区
3. 夏日屋
4. 长仓库
5. 网球场
6. 活动办公室
7. 游戏室
8. MES会议中心
9. 法式滚球场
10. 酒吧露台
11. 游泳池
12. 高尔夫商店

1. Imposing splendour Cathedral Room is an intimate place.
2. The inspiring Cathedral room has oak panelling.

1. 华丽庄重的教堂厅是一个私密空间。
2. 精美绝伦的教堂厅装饰着橡木镶板。

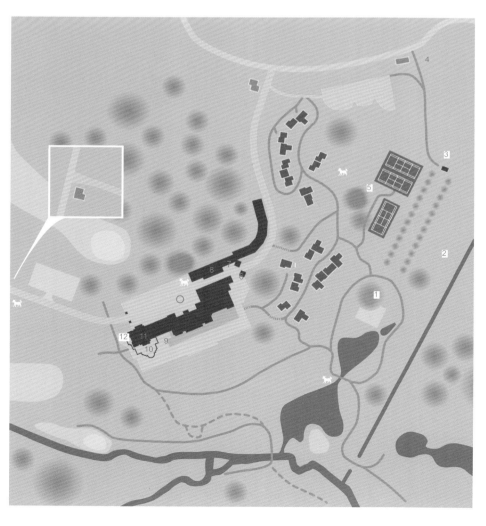

1. The Adam room has stunning period décor, chandeliers and views over the estate.
1.亚当厅拥有优美的复古装饰和吊灯,俯瞰着酒店。

1. The accessible bedroom on the ground floor features an equisite fireplace.
2. The flooring in the accessible bedroom is short pile carpet.
3. The Classic Room has the stylish décor.
4. Superior Castle Room has a generously-sized bedroom space and a compact seating area for private dining or work.

1. 一楼的客房以精致的壁炉为特色。
2. 客房的地面铺有短桩地毯。
3. 古典厅的装饰十分时尚。
4. 顶级城堡套房设有宽敞的卧室和紧凑的休息区（可供私人就餐和工作）。

Crystal Lamp
水晶灯

When Crystal Lamp is not Only Luminary
当灯具不止是光源

The elegant and luxurious neo-classical style can not leave the embellishment of crystal lamps. Resplendent and dazzling crystal lamps inherit the classical essence and absorb the modern beauty, which can be utilised in all the corners of hotel. Besides the function of illumination, the crystal lamp is an important element of the neo-classical style pursuing luxury and grandeur, and can serve as the synonym of "luxury". Based on different functions, the crystal lamp has a rich catalog including desk lamps, pendant lamps, ceiling lamps, wall lamps and so on. Among them, the shapes of crystal pendant lamps are the most intricate, and chandeliers in particular. Chandeliers are adopted most widely, such as in public areas and guestrooms. Crystal ceiling lamps inlayed into the ceiling are usually decorated in big ballrooms. Meanwhile, crystal wall lamps are installed in corridors and other circulation areas.

In Charleston Place where a variety of crystal lamps gather, the lobby is decorated with intricate candle-holder-shaped chandeliers. Around the chandelier are upwards bended candle-holders, under which the part with primitive simplicity and luxury is embellished with crystal, filling the spacious space of lobby perfectly and conforming to the principles of symmetry of neo-classicism rigorously. Moreover, eight huge oblate crystal ceiling lamps decorate the luxurious ballroom. The ceiling lamp is composed of thousands of crystals. Simple as it is in the shape, the ceiling lamp is still dazzlingly colourful and extraordinarily luxurious, shining over 1000m^2 ballroom sumptuously. The costly ceiling lamp is very decorative, and it is usually custom-made and only used by hotels and other high-end venues.

优雅华贵的新古典风格离不开水晶灯的点缀，璀璨耀眼的水晶灯传承着古典的底蕴，也有现代的美感，在酒店中的所有角落几乎都能用到。除了照明之外，它是新古典风格追求豪华大气的重要元素，可以说是奢华的代名词。水晶灯的种类很多，根据功能的不同有水晶台灯、水晶吊灯、水晶吸顶灯、水晶壁灯等。其中，水晶吊灯的造型最多，尤其是枝形吊灯，在酒店中的应用最广泛，公共区和客房都可以使用。水晶吸顶灯是嵌入天花板上的水晶灯，一般用在大型宴会厅中，而水晶壁灯通常会用在酒店走廊等交通区。

查尔斯顿酒店中集合了许多类型的水晶灯，大堂中采用了造型复杂的烛台形枝形吊灯。吊灯四面围绕着弯曲向上的烛台，下面点缀着水晶制成的流苏，造型古朴又华贵，很好地填补了大堂宽敞高大的空间，严谨的设计也贴合了新古典主义的对称原则。而奢华的宴会厅中用了八盏巨大的扁球形水晶吸顶灯，这种吸顶灯由上千块水晶连接而成，造型虽然简单，但流光溢彩，异常华丽，超过1000平方米的宴会厅在这些水晶灯的映衬下极尽豪华。这种大型的水晶吸顶灯造价昂贵，装饰性也很强，通常都需要专门的定制，只在酒店或其他高档场所使用。

Charleston Place

查尔斯顿酒店

Location: Charleston, USA
Designer: Cannon Design
Photographer: Charleston Place
Area: 22,000m²

项目地点：美国，查尔斯顿
设计师：坤龙建筑设计公司
摄影师：查尔斯顿酒店
面积：22,000平方米

A perfect blend of 18th century style and 21st century comfort, Charleston Place is situated in the heart of Charleston, South Carolina, one of America's oldest cities. The hotel is famous for its traditional southern hospitality.

The design of Charleston Place is the feeling of 17th century grandeur with Italian marble lobby with it's Georgian Open Arm staircase to it's 12 ft Venetian crystal chandelier and luxurious suites. The hotel has 440 rooms, including suites and a private Club Level, all the rooms have Botticnio marble and brass bathrooms, all rooms feature period furnishings with Southern style armouries and artwork by local prominent are is reflecting Charleston's past, there are two club floors for members with a number of privileges. (Botticino marble is from Brescia in Italy) Charleston Place is a southern hospitality paradise, with its wonderful shops and Italian marble reception and the "Gone With the Wind Staircase" takes you to another world of southern hospitality.

Charleston is one of America's most preserved architectural and historic histories, over 300-year-olds, church bells toll on the hour as history unfolds around every corner. The English settled in 1680 and King Charles II named the city Charles Town, it became a flourishing seaport. By the early 1800's Charleston was living a second century as one of the most fashionable cities in the New World, it was only second to New York. In April 1861 Confederate soldiers fired on Fort Sumter in Charleston Harbour (which you can visit and relive the battle), thus signalling the beginning of a devastating war, which took Charleston years to recover from.

1. Charleston Grill Restaurant decorated with wooden walls.
2. Thoroughbred club offers a cozy environment.
3. The renwned restaurant has beautiful arched door.

1. 查尔斯顿烤肉餐厅采用了木墙装饰。
2. 优品俱乐部提供了舒适的环境。
3. 这家远近闻名的餐厅采用了美观的拱门。

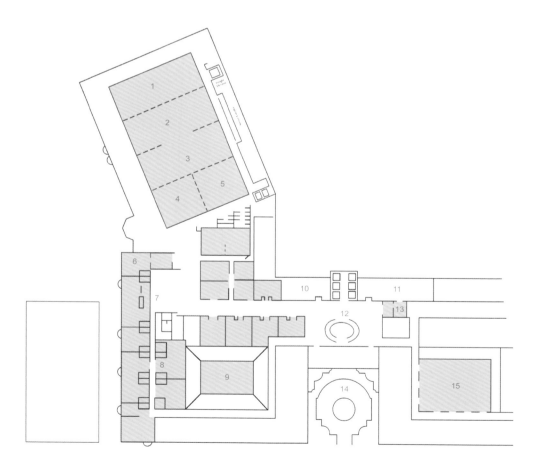

Floor plan
1. Willow
2. Magnolia
3. Live oak
4. Cypress
5. Dogwood
6. Conference lounge
7. Guest corridor
8. Hospitality suites
9. Courtyard Palmetto Café
10. Executive office
11. Sales office
12. Stairs to lobby
13. Business centre
14. Market street circle
15. Pavilion

平面图
1. 柳树厅
2. 木兰厅
3. 槲树厅
4. 柏树厅
5. 山茱萸厅
6. 会议休息室
7. 客房走廊
8. 酒店套房
9. 棕榈庭院咖啡厅
10. 行政办公室
11. 销售办公室
12. 大堂楼梯
13. 商务中心
14. 市场街
15. 休息亭

完美结合18世纪风格和21世纪风尚的查尔斯顿酒店坐落在美国加州南部的查尔斯顿——美国最古老的城市之一。酒店以其传统的美国南方特色而著称。

查尔斯顿酒店的设计展现了17世纪的奢华风格，从由意大利大理石装饰的大堂登上乔治王朝风格的开放式扶手楼梯，从而到达12英尺（约3.66米）高的维也纳水晶吊灯和奢华的套房。

酒店拥有440间客房（包括套房和私人会所层），所有房间的浴室都由大理石和黄铜制品进行装饰，房间内采用复古装饰，点缀着美国南方风格的军械武器。由当地著名人士创造的艺术品反映了查尔斯顿的过去，两层私人会所将为会员提供专属服务。查尔斯顿酒店是南方的度假天堂，酒店内奇妙的店铺、意大利大理石前台以及电影《乱世佳人》中的楼梯将带你领略美国南方的热情好客。

查尔斯顿是美国最传统的历史城市之一，300多年来，教堂的钟声从未停止。英国人在1680年开始在查尔斯顿定居，查理二世国王将其命名为查理镇，使其成为了一个繁华的港口。19世纪初，查尔斯顿是"新世界"（美洲新大陆）最时尚的城市之一，仅次于纽约。1861年4月，联盟士兵在查尔斯顿港的萨姆特堡开火，标志着残酷的美国内战开始。此后，查尔斯顿经过了数年才恢复往日的繁华。

1. A corner of suite
2. An alluring mix of elegance, space and sophisticated grandeur, signature suite offer the ultimate in luxurious accommodation.
3. Beautiful paintings, period antiques and a palette of soothing tones combine to create a powerful, yet refined ambience.
4. Located on the top floor, this sumptuous suite offers an enticing combination of space and privacy

1. 套房一角
2. 极品套房集优雅、宽敞和精致华丽于一身，提供了极致奢华的住宿体验。
3. 美丽的绘画、古董和舒缓的色调共同营造出强大而精致的氛围。
4. 这间位于顶楼的奢华套房巧妙地结合了空间的宽敞度和私密感。

175

Textile
织物

When Softness Becomes a Power
当柔软成为一种力量

The soft textile is of much concern for ornamenting, and is an important element for creating warm and comfortable environment. In hotels, among rich textiles, comfort is the foremost standard for selecting. Secondly, it should be luxurious and elegant matching with the features of the neo-classical style, and can consider the hotel's local features and express its spiritual connotation. The common materials are cotton, velvet, silk and brocade, and patterns are usually bouquets, coloured ribbons, garlands, birds, strip, etc. Moreover, the embroidery from the Orient is very popular. The ornaments for the neo-classical style textiles are lavish, such as curtain tassels, pendants, cushion laces and bed skirts. The ornaments will showcase the decorative effect better and create artistic environment on the condition that they will not affect the function. In the Milestone Hotel and Apartments, every corner is embellished with different textiles, such as calm black printed carpets, retro cotton curtains, festive red sofas, fresh floral pattern chairs and British style plaid bed linens. Different textiles of each guestrooms display different scenery. Sometimes, it is the luxury of silk from the Orient; sometimes, it is magnificent baroque; and sometimes, it is the freshness from British countries. In the hotel, ceilings and walls are not inlayed or gilded too much, and textiles become the main element in the interior environment. The important point is the match between textiles and furniture, between walls and grounds, and between textiles and textiles, which is also the point that can display the design style perfectly.

柔软的织物也是酒店中的重要装饰，是营造温暖舒适环境的重要元素。酒店中所用的织物很多，舒适度是选择的首要条件，其次，还要符合新古典风格华贵典雅的特点，并且还要充分考虑到酒店的地域特点和所要表达的精神内涵。棉布、天鹅绒、丝绸、锦缎是常见的材料。面料的图案常用花束、彩带、花环、鸟类、条纹等，另外，来自东方的刺绣也很受欢迎。新古典风格的织物配饰也有很多，如窗帘的流苏，挂件、垫子的花边、床裙，在不影响功能的前提下，饰品会更好地展现织物的装饰性，营造具有艺术感的环境。里程碑酒店的每个角落都被不同的织物包围着。有沉静的黑色印花地毯、复古的棉布窗帘、喜庆的大红沙发、清新的花朵椅子，也有英伦风格的格子床品。每间客房都有不同的织物搭配，展现了不同的风景，时而是东方丝绸的华贵、时而是巴洛克的恢宏、时而是英国乡村的清新。在酒店中，天花板、墙面并没有过多的镶嵌镀金等处理方式，织物成了主导室内环境的要素，织物与家具的搭配、与墙面和地面的搭配、织物和织物之间的搭配是设计中的重点，也是设计风格展现得最为淋漓尽致的地方。

The Milestone Hotel and Apartments

里程碑酒店

Location: London, UK
Designer: Bea Tollman
Photographer: The Milestone Hotel and Apartments
Area: 3,850m²

项目地点：英国，伦敦
设计师：比·托尔曼
摄影师：里程碑酒店
面积：3,850平方米

With its 12 suites and 45 deluxe guestrooms, The Milestone is intimate enough to offer a personalised service, to have the staff know and care about you. The decor by Bea Tollman is very British, often theme-based and with theatrical flair. Enjoy an afternoon tea at the Park Lounge with the view on Kensington Palace and Gardens, a library with 19th century books and a cosy fireplace. You can have dinner in the intimate Oratory, adjacent to the Cheneston Restaurant. It is the mansion's ex-chapel.

All rooms are unique. The studios and suites are theme based. For instance, chose between the Military, Ascot, Botanical and Margaret Rose studio. Room 605, the Paris Studio, is decorated in black and onyx Luttece fabric. It blends classic design with state of the art technology including pop up television and electronic blinds. The bathroom is equipped with a Jacucci, separate shower and second television set. The Paris Studio's two skylights enhance the room's brightness. Last but not least, it enjoys panoramic views across Kensington Palace and Gardens. The Savile Row room is decorated with pinstripe and a half-finished suit on a tailor's mannequin. The Safari Suite is bamboo bound, features a King bed under a colonial ceiling fan and is decorated with Safari print fabric, wildlife paintings, Africa wicker baskets and ornate ostrich eggs. Comfort is added to this theme suite by an original chesterfield and mahogany furniture. The Safari Suite offers a panoramic view across Kensington Palace and Gardens.

Room 106, the Prince Albert Suite is a spacious high ceiling suite with original leaded Victorian windows, again overlooking Kensington Gardens and Palace. It features a king size four-poster bed and an elegant seating area including a chaise longue. The large bathroom has a separate shower, Jacucci and television. The Prince Albert Suite's very special feature is its ornate balcony, perfect for entertaining.

Rom 208, The Club Suite, is a duplex Grand Suite with unique floor to ceiling windows overlooking Kensington Palace and Gardens. The upstairs salon features an antique billiard table, a spacious seating area and ensuite restroom. The red and gold walls and leopard print carpet add a touch of luxury and comfort to The Club Suite.

里程碑酒店拥有12间套房和45间豪华客房,酒店的员工将为宾客提供无微不至的专属服务。比·托尔曼的装饰英式特色突出,常常具有主题性和戏剧感。宾客们可以在公园大堂酒吧边享用下午茶边观赏肯辛顿宫和花园的美景,或者在图书室阅读19世纪的古书享受舒适的壁炉,还可以在查奈斯顿餐厅旁的宣讲餐厅(酒店的前礼拜堂)中享用美食。

所有客房都别具一格。工作室和套房都有自己的主题,如军事工作室、阿斯科特工作室、植物工作室、玛格丽特·罗丝工作室等。605房的巴黎工作室采用黑色和黑玛瑙织物进行装饰,结合了古典设计风格和现代化科技(包括弹出式电视和电动百叶窗)。浴室内配有极可意按摩浴缸、独立淋浴间和电视设备。巴黎

工作室的两扇天窗提升了房间的明亮度。重要的是,它还享有肯辛顿宫和花园的美景。

萨维尔街客房采用细条纹装饰,屋内摆放着一件未完成的西装。狩猎套房以竹子包围,殖民地式吊扇下是特大的双人床。房间采用印花布、野生动物绘画、非洲柳条篮和华丽的鸵鸟蛋进行装饰。定制的躺椅和桃心木家具为房间增添了舒适感。狩猎套房同样享有肯辛顿宫和花园的美景。

106房是阿尔伯特亲王套房,宽敞高大的套房配有维多利亚式窗户,房间也俯瞰着肯辛顿宫和花园的美景。套房配有特大号四柱床,优雅的休息区包含一张躺椅。大浴室内配有独立淋浴间、按摩浴缸和电视。阿尔伯特亲王套房的特色在于它的装饰阳台,十分适合客人在阳台上娱乐。

208房是独占两层的会所套房,其特有的落地窗俯瞰着肯辛顿宫和花园的美景。楼上的沙龙配有古董台球桌、宽敞的休息区和套房洗手间。红色和金色的墙壁和豹纹地毯为套房增添了奢华感和舒适度。

1. Small but warming reception
2. The beautifully decorated function room is fresco-style.
3. The conservator is blace and white, but it's very bright.

1. 小而温馨的酒店前台。
2. 装潢精美的功能厅采用了壁画风格。
3. 等候区采用黑白装饰,十分明亮。

1. Park lounge is also a small library.
2. Intimate dining area with beautiful fireplace
3. Gorgeous restaurant with oriental style.

1. 公园酒吧同时还是一间小型图书室。
2. 设有壁炉的私密就餐区。
3. 东方风格的优雅餐厅。

1. Seperated dining room and sitting room in suite
2. Recently refurbished suite has stunning cream and beige colour scheme.
3. The self-contained apartments opposite Kensington Palace, all of which are individually designed, it combines all the luxuries of the hotel with extra space and privacy.
4. A corner of Junior Suite
5. Detail of Master Suite

1. 套房的独立餐厅和客厅。
2. 重新装修的套房以奶白和浅黄为主色调。
3. 设施齐全的公寓位于肯辛顿宫对面，全部采用独立设计，兼具酒店的奢华和住宅的私密。
4. 普通套房一角。
5. 主人套房的细部设计。

1. Harlequin Suite is uniquely decorated with black fabrics and furnishings.
2. The Living Room of William and Catherine' Suite
3. The Living Room of Edwardian Suite
4. A Deluxe King bedroom has warm a style.

1.斑斓套房装饰着独一无二的黑色织物和家具。
2.威廉姆和凯瑟琳套房客厅。
3.爱德华套房客厅。
4.豪华国王卧室采用了温暖的风格。

1. In Junior Suite guest can be decadently modern, unashamedly homely, or elegantly traditional.
2. Deluxe King Room full of opulent fabrics and furnishings.
3. A Superior Queen room recalls the time when luxury and courtesy were a way of life for wealthy Londoners. This is a queen-bedded room filled with fine fabrics and antique furnishings.
4. A series of stunning decorative schemes bring each Deluxe Studio to life.
5. Master Suite is a spectacular one-off, an ornate adventure in five-star luxury.

1. 普通套房的客人可以现代时尚，可以居家简单，也可以优雅传统。
2. 豪华国王房满是奢华的织物和装饰。
3. 豪华皇后房令人回想起伦敦人曾经奢侈且彬彬有礼的生活方式；大床房内布满了精致的织物和古董家具。
4. 一系列华丽的装饰让豪华工作室充满了生活气息。
5. 主人套房将为宾客提供华丽的五星级奢华体验。

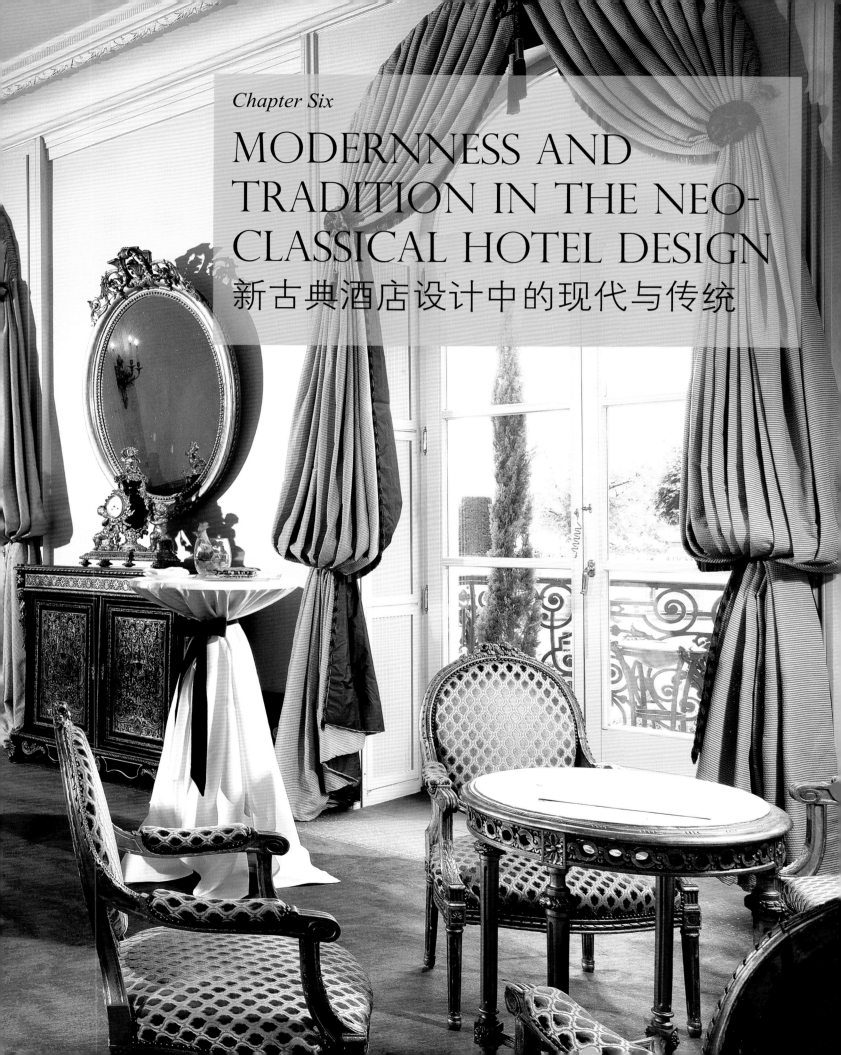

Chapter Six

MODERNNESS AND TRADITION IN THE NEO-CLASSICAL HOTEL DESIGN
新古典酒店设计中的现代与传统

Chapter Six

MODERNNESS AND TRADITION IN THE NEO-CLASSICAL HOTEL DESIGN
新古典酒店设计中的现代与传统

The neo-classical style emerged in the 18th century. The excavation of the ancient Rome and Greece relics started the exploration and study on the ancient world. The neo-classical style absorbed some elements from the relics, aiming to create the balance and delicate space. Its presentations are diverse: in Britain, it is called the Georgia style, the federation style in the US and the Louis XVI style in France. The art form highlights reasonable proportion and elegant disposition with classical details, which has been popular for 100 years.

新古典风格产生于18世纪，古希腊、古罗马遗迹的发掘开启了人们对古代世界的探寻和学习，新古典风格汲取了其中的一些元素作为装饰，旨在创造出均衡、精致的空间。新古典风格的表现形式有很多，在英国称为乔治亚风格，美国称为联邦风格，法国称为路易十六风格。这种艺术形式强调合理的比例和气质的优雅，装饰有古典的细节，在西方流行了100多年。

19世纪的工业革命以后，建筑材料和结构有了重大的改变，钢铁、玻璃等材料代替了原来的石材、木头，建筑的形式也发生了改变。随之而来的，是室内设计也发生了变化，墙面不再只限于镶板的装饰，各种类型的墙纸使用方便并且也有很强的质感；纺织技术的进步使大规模的地毯、挂毯等织物的生产更加方便，地面也不再只有拼花地板；装饰材料的改变促使着设计风格有了新

The Wintergarten Restaurant, flooded with natural light.

花园餐厅充满着自然光。

Chapter Six

MODERNNESS AND TRADITION IN THE NEO-CLASSICAL HOTEL DESIGN

After the industrial revolution in the 19th century, tremendous changes happened on the architectural structure and materials. Steel and glass replaced stone and wood, and so did the architectural forms. Consequently, the interior design changed – walls were not only decorated by panels because various wallpapers with excellent texture were convenient for use; the textile technology was improved, making the mass production of carpets, tapestries and other textiles more convenient and thus, the ground was not floored with the parquet flooring any more. Such changes in decorative materials impelled the design style to develop and neo-classicism has new forms gradually.

Besides materials, the development of modern equipment technology dramatically transformed the interior design style. In the mid-19th century, Elisha Graves Otis, an American inventor, showed the first elevator in World Exposition of New York; at the beginning of the 20th century, a department store in Detroit was equipped with 3 central air-conditioners, opening the era of air-conditioning serving people; in 1924, British engineer John Baird invented the first television which greatly influenced the lifestyle and the information dissemination. These inventions provide comfort and convenience to people's life. Then, these facilities were firstly used in the hotel, the commercial space. However, in the meanwhile, new facilities also brought big impact on the traditional interior design, which not only changed the space shape, but changed the organisational functions in the interior space on the basis of the requirements of the circuit layout, ventilation device installation, water circulation, etc. Besides luminaries, air-vents, signal amplifiers, rainmakers, horns, etc. which were set on the ceilings, the furniture, televisions, telephones, faxes and other modern facilities were also installed in the room. To the neo-classical style deep rooted in rigorous proportion and classical decoration, it was a huge challenge not to have these facilities excluded. As the invention of computer in

的发展，新古典主义也渐渐有了新的形式。

除了材料之外，现代设备技术的发展也大大地改变了室内的设计风格。19世纪中期，美国发明家伊莱沙·格雷夫斯·奥的斯（Elisha Graves Otis）在纽约的世界博览会上向人们展示了第一部升降梯；20世纪初，底特律一家商场安装了三台中央空调，开始了空调为人服务的时代；1924年，英国工程师约翰·贝尔德（John Baird）发明了第一台电视机，对人们的生活方式、信息传播产生了重大影响。这些设备的发明为人类生活带了舒适和便捷，于是，酒店这种商业空间开始率先使用。但与此同时，新的设备对传统的室内空间设计也带来了巨大的冲击，不仅室内的空间形态有了变化，电路的铺设、通风设备的安装、水循环的要求等也改变了室内的组织功能。天花板上除了灯具之外，还有通风口、信号放大器、喷淋设备、喇叭等，室内除了家具之外，也有了电视、电话、传真机等现代设备。对崇尚严谨比例和古典装饰的新古典风格来说，这些设施的加入是一个巨大的挑战。

随着1946年电子计算机的发明，人类开始了新一轮的信息技术革命，计算机的普及、互联网的使用、现代通讯技术的发展使高科技开始走入人们的生活，同时，绿色节能的理念开始受到越来越多人的接受，室内的设计也开始了智能化和生态化的历程。相对于简洁的现代风格来说，新古典风格需要更多的装饰元素，如何在复杂的装饰中保证现代酒店的舒适、智能、环保要求是设计者共同的思考。

自新古典主义风格产生以来，已经经历了二百多

1946, human began a new revolution – the information and technology revolution. The popularization of computers, the application of the Internet and the development of the modern communication technology gave high technology access to people's life. In the meanwhile, the concept of green energy started to be accepted by more and more people and the interior design also commences its journey to intelligentialise and ecologicalise. Compared with the simple modern style, the neo-classical style requires more decor elements. And how to assure modern hotels comfortable, intelligent, and environmental under the circumstance of intricate decorations is what the designers should think over.

Since the day when it was born, the neo-classical style has experienced over 200 years during which it witnessed great changes took place in human's life. The interior design has experienced countless transformations in styles. However, neo-classicism is still loved by many people, becoming the first choice in high-end hotels. That is, the vitality of neo-classicism is more or less related to its tolerance of modern forms and the harmony of modern facilities.

年的历史，人类的生活在这二百多年中发生了翻天覆地的变化，室内设计也历经了不计其数的风格变迁，但新古典主义依然被很多人所喜爱，成为高档酒店中的首选，可以说，这种鲜活的生命力与其对现代形式的包容和与现代设施的和谐不无关系。

Beau-Rivage, set in splendid historic surroundings, has salons that can cater for up to 350 guests.

美岸酒店拥有璀璨的历史背景，酒店内的沙龙最多可容纳350人。

Coordination between Traditional Designs and Modern Facilities
传统设计与现代设施的协调

The Collide between Baroque Decoration and Modern Technology
巴洛克装饰与现代技术的碰撞

The requirements of comfort in modern hotel become higher and higher and more and more facilities are needed. For the neo-classical interior design, these indispensable facilities are not little challenges. From television, computer and air-conditioner to each outlet, if designed inappropriately, it will make visual conflict with the neo-classical style in the interior, or will cause the loss in function. In the progress of design, the designer usually adopts the way of hiding and transformation to tackle the above problems. For example, some renowned hotels hide televisions above fireplaces, and cover them with panels or mirrors. And some outlets are used as ornaments in accordance with the interior environment and walls. The principle of the methods of hiding and transformation is not to affect the function and integrate with the surrounding environment. Situated in Geneva, Hotel Beau Rivage, with a long history, is decorated delicately, among which some suites showcase the shadow of the baroque style. Though decorated classically, it is equipped with the most advanced facilities and smart designs. In the bathroom of Imperial Suite, Jacuzzi is designed with four orders around the bathtub and a Swan-shaped faucet is in accordance with the design style. The salon with capacity of 350 persons provides state-of-the-art facilities for meeting the needs of various types of conferences. However, these facilities hiding behind the neo-classical appearance do not affect the entire style of the hotel. Here, even a little container is carved delicately without any modern brands.

现代酒店对舒适度的要求越来越高，现代化的设备也越来越多，这些不可缺少的设备对于新古典的室内装饰来说是个不小的挑战，大到电视机、电脑、空调，小到一个电源插座，如果处理不当，就会与室内的新古典风格产生视觉上的冲突，或是产生功能上的缺失。在设计中一般会用隐藏和转化的方式解决此类问题。如一些知名酒店会把电视机隐藏在壁炉上方，用镶板盖住，或是隐藏在镜子后面。而一些电源开关等会做成与室内环境一致的装饰品，与墙面统一。这些隐藏和转化的原则是不影响其使用功能，又能恰当地和周围环境融合。

位于日内瓦的美岸酒店（Hotel Beau Rivage, Geneve）历史悠久，装饰优雅，一些套房中还带有巴洛克风格的影子。尽管装饰古典，但这里却有着最先进的设备和巧妙的设计。皇家套房的浴室中带有豪华的按摩浴缸，为了保证设计风格的一致，浴缸四周设计了四根柱式，连水龙头也被设计成了天鹅形。能容纳350人的会议沙龙提供了最先进的技术设备，能满足各种类型的会议需求，而这些设备完全被隐藏在新古典的装饰下，丝毫不会影响酒店的整体风格，在这里，即使是小小的储物盒也精心雕刻，看不到现代的烙印。

Hotel Beau Rivage, Geneve

日内瓦美岸酒店

Location: Geneva, Switzerland
Designer: Ms. Leila Corbett
Photographer: Hotel Beau Rivage, Geneve
Area: 5,500m²

项目地点：瑞士，日内瓦
设计师：莉拉·科尔比特
摄影师：日内瓦美岸酒店
项目面积：5,500平方米

The Hotel Beau Rivage is a historic 5-star Hotel with 93 rooms. It is located on the shores of Lake Geneva in the very heart of Geneva.

The facilities of the Hotel Beau Rivage include two restaurants: the "Le Chat-Botte" for French Gastronomic cuisine and the "Le Patara" for lovers of exotic and spicy dishes; and the Bar "L'Atrium" for light snacks, afternoon tea, cocktails and piano entertainment in the evening (with open terrace during summer months). Further amenities at the Hotel Beau Rivage include various conference and banquet rooms.

The suites have all retained the outstanding décors and spirit of the 19th century, and have been lovingly restored over the years. All the suites include a bedroom with an adjoining living room (minimum 80 square metres). The windows provide a dress circle seat to enjoy the superb panorama of the lake with the "Jet d'Eau"-fountain, the Alps with the "Mont-Blanc" mountain, and the city. The frescoes by Italian artists of the past century and the original panelling decorate these exceptional apartments. The suites offer several possible combinations from two to seven bedrooms; the most prestigious combination is called "Royal" such as the "Louis II of Bavaria" Suite with up to seven bedrooms leading off from a large and lavish reception room dating from the period of this madly eccentric king. The master bedroom contains an impressive blend of baroque charm and modern technology, with two bathrooms in precious marble – one with a Jacuzzi offering a view of the lake and the Alps; the other with a Sauna and a shower. The absolutely superb décor is in shades of indigo and sienna, with furniture dating from 1847, a crystal chandelier and two queen-size beds.

五星级酒店美岸酒店拥有93间客房，位于日内瓦湖畔，处在日内瓦市的中心区。

美岸酒店拥有两间餐厅：查特波特法式餐厅和帕塔拉异域餐厅；同时，酒店还设有各种各样的会议厅和宴会厅。

酒店套房保持了19世纪以来的杰出装饰和氛围，重新进行了修复。所有套房都设有卧室和客厅（最小面积为80平方米）。从窗口的圆形座位可以看到日内瓦湖的"水蒸气"喷泉、阿尔卑斯山的"勃朗峰"和城市的美景。由意大利艺术家制作的壁画和原始镶板装饰让客房异常美丽。套房的卧室从二间到七间不一，最知名的是皇室组合：路易维希二世套房拥有七间卧室，宽敞奢华的接待室让人想起了这位古怪无常的国王。主卧室结合了巴洛克装饰和现代技术，两间浴室都采用珍贵的大理石装饰 —— 一个带有按摩浴缸，可以享受湖泊和山脉的美景，另一个带有桑拿房和淋浴间。极致奢华的装饰以靛蓝色和赭石色为主，配有源于1847的家具、水晶吊灯和两张超大尺寸的大床。

1. L'Atrium Bar is a must for those in the know in Geneva, with its gorgeously hushed surroundings, a perfect escape from the hustle and bustle of the city.
2. Patara Restaurant features lavender colour.

1. 中庭酒吧是日内瓦的必去之处，人们可以在那里远离周围城市的喧嚣。
2. 帕塔拉餐厅以淡紫色为主。

Floor plan
1. De Brunswick salon
2. Windsor salon
3. Imperatrice salon
4. Tremois salon
5. Foyer
6. Sarah bernhardt salon
7. Masaryk salon

平面图
1. 不伦瑞克沙龙
2. 温莎沙龙
3. 皇后沙龙
4. 特莱莫沙龙
5. 门厅
6. 贝恩哈特沙龙
7. 马萨里克沙龙

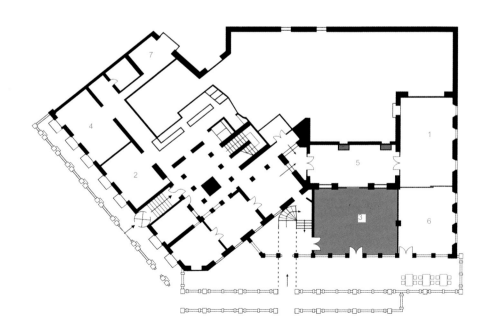

1. Sissi suite has a very spacious living room.
2. The salon's size makes it suitable for fashion shows and art exhibitions. Its stunning surroundings are also ideal for those seeking.
3. This room was designed and decorated in the 1970s as a tribute to the famous French painter, sculptor and engraver Pierre-Yves Trémois.

1. 茜茜套房拥有宽敞的客厅。
2. 沙龙的规模适合举行时装秀和艺术展览。
3. 房间设计采用了20世纪70年代的风格，旨在向法国著名画家、雕塑家和雕刻师皮埃尔-伊夫·特莱莫斯致敬。

1. Marble fireplace and red sofas in lobby
2. Six historic suites are imbued with the spirit of a private abode. Even more than the others, they have retained traces of their original décor, their wood panelling and their fine Italian frescoes.

1. 大堂的大理石壁炉和红色沙发。
2. 六间历史套房充满了私人住所的气息；它们更好地保留了原始设计、木质镶板和精致的意大利壁画。

1. This magnificently restored suite is masterpieces of harmony.
2. Combining creativity and modernity, the Imperia Suite and the Royal Suite offer comfort and peace in a refined and elegant setting.
3. In keeping with its tradition of excellent hospitality, the most intimate of the great Geneva hotels offers the very best in the art of living.
4. With their balconies looking out over the spectacular scenery of Lake Geneva, these suites have long been a source of inspiration.
5. This magnificently restored suite is a masterpiece of harmony which gently murmurs the names of Sissi, Wagner and Cocteau – all distinguished guests at Beau-Rivage.

1. 这间经过修复的宏伟套房是和谐设计的典范。
2. 统帅套房和皇室套房结合了新颖和现代化的装饰，在精致优雅的环境中为宾客提供了舒适和宁静。
3. 日内瓦美岸酒店继承了优秀的服务传统，展现了最佳的生活艺术。
4. 套房的阳台可以俯瞰日内瓦湖的壮丽景色，一直是人们的灵感来源。
5. 这间华丽的套房曾经居住过茜茜公主、瓦格纳和科多拓等名人。

Application of Modern Materials in Traditional Environment
现代材料在传统环境中的应用

Vivid Classical Style
活力古典风格

The finishing materials are abundant in the hotel design. According to the different parts needing decorations, the materials can be classified as ceiling finishing materials, wall finishing materials and ground finishing materials. The ceiling finishing materials generally include plastic suspended ceiling, glass ceiling, wooden decorative sheet, parquet decorative sheet, metal ceiling, etc. Wall finishing materials include wall paint, wallpaper, wooden decorative sheet, metal decorative sheet, fresco, stone veneer, wall tile, etc. Ground finishing materials include ground coating, wooden floor, bamboo floor, man-made floor, ground tile, carpet, etc. The neo-classical style has strict requirements for the texture, size and gloss. Not all the materials are suitable to the style. Wooden and plaster suspended ceilings are easy to carved, which are the common ceiling materials. Fresco and wooden decorative sheets are traditional wall finishing materials, and today, wallpapers are more popular in hotels, especially natural silk wallpaper, textile wallpaper and so on, which are luxurious, easy to install, and more environmental than other traditional ones. As to ground finishing materials, solid wood parquet is replaced by modern laminate flooring with carpets.

Brenners Park-Hotel & Spa is a hotel combining both modern and classical styles, which is called "vivid classical style" by its designer. There are not only antique fireplaces and furniture, but modern coated walls and glass walls; there are not only traditional silk and linen textiles, but modern glass bar counters; there are not only British style suites, but fashionable outdoor restaurants. Traditional European decorations are reserved in details, and the application of modern materials assures the comfort of hotel.

酒店的装饰材料很多，按照装饰部位可分为顶棚装饰材料、墙面装饰材料和地面装饰材料。顶棚材料一般有塑料吊顶、玻璃吊顶、木制装饰板、石膏装饰板、金属吊顶等。墙面材料包括墙面漆、墙纸、木制装饰板、金属装饰板、壁画、石饰面板、墙面砖等。地面材料有地面涂料、木制地板、竹地板、人造板地板、地面砖、地毯等。新古典风格对材料的质感、规格、光泽度等要求较高，并不适用于所有材料。木制吊顶和石膏吊顶易于雕刻，是常用的顶棚材料，壁画和木制装饰板是传统的墙面材料，如今，墙纸更容易被酒店所接受，尤其是天然材料的丝绸壁纸、纺织物壁纸等，效果华丽并且方便安装，较之于传统的方式更环保。而地面材料则从原来的实木拼花地板转变为现代的复合地板加地毯的方式。

布莱纳斯公园酒店是一家融合了新与旧的酒店，设计师称其为"活力古典风格"。这里不仅有古老的壁炉和家具，也有现代的涂漆和玻璃墙面；有传统的丝绸和亚麻织物，也有现代的玻璃吧台；有英伦风格的套房，也有时尚的户外餐厅。细节处保留了欧洲传统装饰的同时，也充分应用了现代材料保证酒店的舒适。

Brenners Park-Hotel & Spa

布莱纳斯公园酒店

Location: Freiburg, Germany
Designer: Countess Bergit Douglas
Photographer: LHW
Area: 5,000m²

项目地点：德国，弗莱堡
设计师：波捷特·道格拉斯伯爵夫人
摄影师：LHW
项目面积：5,000平方米

Brenners Park-Hotel & Spa is only a 10-minute walk to the center of the Old Town, the world famous Belle Epoque Casino, the theatre, the Museum Frieder Burda and the Kulturhaus LA8.
The Foyer, as the entrance to the Orangerie, opens into a luxurious setting, enhanced by priceless Gobelin tapestries and a historic fireplace. Reflecting Countess Douglas' mix of the classic and modern are features such as a small international library and modern bar made of red glass.

Under the leadership of interior designer Countess Bergit Douglas Brenners Park-Hotel & Spa has debuted four new suites. These new accommodations, combining classic architecture and European décor with contemporary touches, maintain the history and grandiosity of the property while offering the modern, luxurious amenities that appeal to today's travelers. The spacious new one-bedroom suites with large windows allow for abundant daylight from the west and south. Each generously proportioned suite has been richly appointed to preserve the values of European home décor. Incorporating Countess Douglas' passion for travel, the suites feature sophisticated furniture and fixtures from France and England, as well as in one case Indian Colonial stitching on sofa, ottoman and bedroom fabrics. The design emphasises strong colours such as red and green, as well as contrasting materials including linen and silk.

"I would classify my style as 'classic with pep,'" said Countess Douglas. "The goal is to bring a sense of living in comfort with style, bound by tradition but with modern elements. Just like an elegant private house, the suites will mix old with new, putting a modern spin on the historic hotel."

211

1. Deluxe Room offer exceptional levels of living comfort, boasting an average size of 50 square metres and are furnished in an outstanding design.
2. The spacious lounge area has a balcony overlooking Baden-Baden's famous avenue.

1. 豪华套房异常舒适，平均面积可达50平方米，其装潢设计也极为出色。
2. 宽敞的休息区所设的阳台可以俯瞰巴登巴登城著名的林荫大道。

布莱纳斯公园酒店距离旧城区的中心、闻名世界的好时代赌场、剧院、布尔达博物馆和LA8文化中心等景点仅有10分钟的步行路程。

酒店的门厅通往内部富丽堂皇的布景，价值连城的哥白尼双面挂毯和古老的壁炉增添了空间的奢华感。道格拉斯伯爵夫人结合了古典与现代设计，增添了一个小型国际图书室和现代红玻璃吧台。

在室内设计师波捷特·道格拉斯伯爵夫人的带领下，布莱纳斯公园酒店新添加了四间套房。这些新套房在古典建筑风格和欧洲装饰中增添了现代气息，保持了酒店的历史和宏大风格，同时也提供了适合当前旅行者的现代奢华设施。

宽敞单卧套房内，巨大的玻璃窗让充足的光线从西面和南面进入室内。每间比例相称的套房都保留了欧洲家居装饰的价值。套房融入了道格拉斯伯爵夫人对旅行的热情，以来自法国和英格兰的精致家具以及运用在沙发、脚凳和床品上的印度殖民地刺绣为特色。设计突出了红色、绿色等浓重的色彩以及亚麻和丝绸等具有对比价值的材料。

道格拉斯伯爵夫人称："我将自己的风格归类于'活力古典风格'。目标是打造舒适时尚的生活，在传统中融入现代元素。正如优雅的私人住宅一样，套房结合了新与旧，在历史酒店中增添了现代特色。"

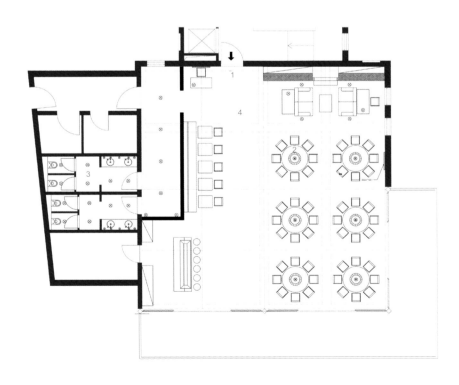

Lounge plan
1. Entrance
2. Ballroom
3. Bathroom
4. Corridor

休息大厅平面图
1. 入口
2. 宴会厅
3. 洗手间
4. 走廊

1. The Salon Minerva, with its light design and furnishings, is particularly suited for board meetings and small conferences with a maximum of 18 delegates.
2. Deluxe Suite B with amazing colours boasts exceptional luxurious design.
3. Imposing alcoves divide the sleeping area from the spacious living quarters in Junior Suite. Thanks to the English inspired design, the Junior Suites provide their very own brand of flair.
4. Superior Room offers a generous living area for greater comfort.

1. 装饰简约的密涅瓦沙龙特别适合举办董事会议和小型会议，最多可容纳18人。
2. B豪华套房色彩独特，拥有十分奢华的设计。
3. 大壁龛将普通套房的卧室和客厅分隔开；英式风格设计让普通套房拥有了独特的品味。
4. 豪华客房提供了宽敞的生活空间和舒适的感受。

1. Elegant décor, every comfort and up-to-date technology combined with attentive, personalised service.
2. Deluxe Suite A combines convenience with exquisite luxury.
3. Each Suite with its fine furniture and small lounge.
4. Rooms in this category offer the high standard of comfort associated with the 5-star superior hotel close to the Black Forest.

1. 优雅的装饰、舒适的空间和先进的技术与贴心的个性化服务结合在一起。
2. A豪华套房既便利又奢华。
3. 每间套房都配有精美的家具和小型休息室。
4. 客房将良好的舒适度和五星级酒店体验结合起来。

Renovation of Historical Sites
历史遗迹的修复

Frescos in the 19th Century Revitalise

19世纪壁画焕发生机

In Europe, many hotels are renovated from antique buildings, some precious relics of which were retained in the progress of renovation, and then formed unique scenery in the hotel. And the domes are the places where frescos are retained most. In European public buildings, domes are usually the essence. From the 15th century, the Renaissance started in Europe centred with Italy and the architecture, the painting and the sculpture were closely connected. Then a great number of sculptors and painters have appeared over the period, whose works enhanced spiritual value of architecture, at the same time, left precious heritage for descendants. The restoration of frescos in the hotel not only retained the original appearance of buildings, but displayed long history. The dome fresco is usually used in ballrooms, lobbies and other high and spacious places, of which magnificence will attract your attention at first sight. The relics of Fairmont Le Montreux Palace can date back to the 19th century. In 1906, it was opened as a brand new hotel. In its history, it experienced several renovations. In the progress of renovation, the designer retained its frescos and stained glass domes in its original ballroom. The walls around and the interior space were designed with intricate carvings and orders to embody the grandeur of the baroque style. In this magnificent ballroom, the original baroque decorations were granted new vitality. The colourful glass domes shine like crystal in the backdrop of chandeliers, around which portraits and arches interweave together. The frescos of a century ago have been reinterpreted in the modern society.

欧洲有许多酒店是由古建筑改造而成的，这些建筑中的一些珍贵遗迹会在翻新过程中得以保留，形成酒店的独特风景。其中穹顶壁画是被保留最多的地方。在欧洲的公共建筑中，穹顶往往是其中的精华所在，自15世纪开始，以意大利为中心的欧洲地区开始了文艺复兴运动，建筑、绘画和雕刻从此形成了密不可分的关系，自此欧洲也陆续出现了许多伟大的雕刻家、画家，他们的雕刻作品为建筑增添了精神价值的同时也为后人留下了宝贵的遗产。酒店对这些壁画进行简单的翻修，不仅保留了建筑的原貌，也体现了其悠久的历史。穹顶壁画这种形式一般会用在宴会厅、大堂等高大宽敞的地方。其壮丽恢宏的画面往往会在第一时间引起人们的注意。

费尔蒙特蒙特勒皇宫酒店的旧址可以追溯到19世纪，1906年，它作为一家全新的酒店正式开业，期间经历了若干次翻新。在翻新过程中，设计师保留了原本宴会厅中的壁画和彩色玻璃穹顶。周围的墙面和室内空间也设计了复杂的雕刻和柱式，以体现新巴洛克风格的壮观。在这间气势恢宏的宴会厅中，原始的巴洛克装饰被赋予了新的活力，彩色玻璃穹顶在水晶灯的映衬下色彩斑斓，周围的人像雕刻与拱形门交织在一起，百年前的壁画在现代社会得到了重新诠释。

Fairmont Le Montreux Palace

费尔蒙特蒙特勒皇宫酒店

Location: Montreux, Switzerland
Designer: Fiona Thomson
Photographer: Fairmont Hotels&Resorts
Area: 16,000m²

项目地点：瑞士，蒙特勒
设计师：费昂纳·托姆森
摄影师：费尔蒙特酒店集团
项目面积：16,000平方米

Fairmont Le Montreux Palace resort hotel is situated on the shores of Lake Geneva in Switzerland, overlooking the Alps,- originally built in 1906, unveil a multimillion-dollar renovation by noted interior designer Fiona Thompson, including a new suite fashioned in honor of flamboyant Queen frontman Freddie Mercury. The ground-floor Grand Hall, with its marble columns, chandeliers, and plush furnishings acts as an extended lobby, is readily converted into a glittering reception hall. It is complemented by six other grandiose spaces bristling with stucco statuary, heavy embroidered drapes, and mahogany furnishings, and a series of more modest conference spaces. Across the road, a further option is La Coupole, with its broad arches and lakeside location.

The hotel's renovation includes 12 conference and meeting rooms, and the 21,000 square foot Willow Stream Spa in addition to the resort's renovated Brasserie, a new lobby lounge and bar, and the famous Harry's New York bar. Elegant and timeless, the guest rooms at Fairmont Le Montreux Palace contrast traditional and contemporary new design elements with breathtaking views of Lake Geneva. Colour palettes are light and airy set against a series of colourful photographic prints throughout the rooms, representing the reflective qualities of water.

The luxurious Beaux-Arts hotel, is introducing a contemporary new redesign of its guestrooms, Brasserie and public spaces, including the Freddie Mercury Suite. Mercury was a Montreux resident and frequent guest of the hotel, as were such luminaries as Tchaikovsky, replica handbags, Stravinsky, Lord Byron, Leo Tolstoy, Dostoyevsky, and Vladimir Nabokov.

The 235-room luxury resort offers 190 elegantly-appointed rooms and 45 suites and junior suites, Elegant and timeless, the guest rooms at Fairmont Le Montreux Palace contrast traditional and contemporary new design elements with breathtaking views of Lake Geneva. Colour palettes are light and airy set against a series of colourful photographic prints throughout the rooms, representing the reflective qualities of water.

费尔蒙特蒙特勒皇宫酒店位于瑞士日内瓦湖畔，远眺着阿尔卑斯山，始建于1906年。著名设计师费昂纳·托普森对其进行了耗资数百万美元的翻新，其中包括一间纪念皇后乐队主唱费雷迪·麦考利的全新套房。麦考利居住在蒙特勒，是酒店的常客。

采用大理石柱、吊灯和长毛绒家具的一层大厅是酒店大堂的延伸，被改造成了金碧辉煌的接待厅。酒店还设有其他六个富丽堂皇的空间（它们内部充满了石膏雕塑、厚重的刺绣帷幔和红木家具）和一系列低调的会议空间。

酒店翻新包括12间会议室、1,950平方米的柳溪温泉水疗会所、全新的百事丽大堂酒吧和著名的哈里纽约吧。酒店的235间房间包括190间客房和45间套房。优雅而经典的客房融合了传统与现代设计元素，远眺着日内瓦湖的美景。客房设计采用轻盈的浅色与鲜艳的形成对比，呈现了水面倒影的效果。

1. Entrance of the hotel
2. Exterior of the hotel
3. Beautiful restaurant close to the big window, it's full of sunshine.

1. 酒店大门。
2. 酒店外观。
3. 靠近大窗的餐厅内充满了阳光。

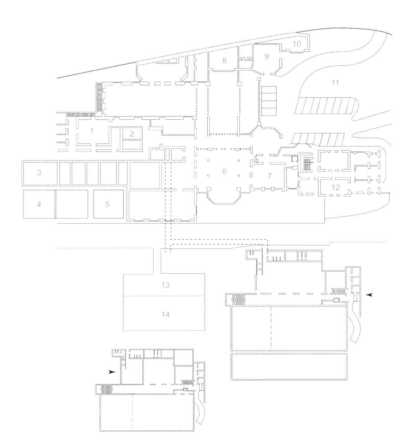

Floor plan

1. Salon rotary
2. Lift
3. Business office
4. Harry's bar
5. Reception
6. Lobby
7. Salon de musique
8. Stage
9. Salon club
10. Salon rouge
11. Garden
12. Salon vert
13. Parking
14. Willow stream spa

平面图

1. 旋转沙龙
2. 电梯
3. 商务办公室
4. 哈里酒吧
5. 前台
6. 大堂
7. 音乐沙龙
8. 舞台
9. 俱乐部沙龙
10. 红色沙龙
11. 花园
12. 绿色沙龙
13. 停车场
14. 柳溪温泉水疗会所

1. The ballroom has Baroque style.
2. The opulent granite staircase and oak-wood floor add to the nostalgic atmosphere of Masson.

1. 巴洛克风格宴会厅。
2. 华丽的花岗岩楼梯和橡木地板增添了空间的怀旧氛围。

1. La Brasserie du Palace decorated in Belle Epoque style.
2. The Lobby Lounge Bar is contiguous to La Brasserie du Palace and offers a direct access to the reception
3. Léman AB is a vast, bright and luxurious space for conferences, product launches and automobile shows.

1. 皇宫酒吧的装饰采用了美好时代（第一次世界大战前）的设计风格。
2. 大堂酒吧是皇宫酒吧的延续，直接与酒店前台相连。
3. 莱蒙AB厅宽敞明亮而又豪华，适合举办会议、产品推介会和车展。

1. The well-established spa of Fairmont Le Montreux Palace is idyllically situated in the gardens of the Palace with a stunning view of Lake Geneva and the Alps.
2. The Palace Lake View Suite is located on the 5th floor and offers one bedroom and a separate living room.
3. The Signature Lake View Suites are spacious suites with bedroom and separate living room with an elegant and luxurious contemporary design.
4. The Bellevue Lake View Suites offer one bedroom and a separate living room in a luxurious and elegant décor.
5. The bed detail of Signature Suite

1. 酒店内精美的水疗中心坐落在花园之中，享有日内瓦湖和阿尔卑斯山的绚丽美景。
2. 皇宫湖景套房位于酒店六层，拥有一间卧室和独立的客厅。
3. 顶级湖景套房十分宽敞，配有卧室和独立的客厅，采用优雅奢华的现代装饰设计。
4. 贝尔维尤湖景套房配有卧室和独立的客厅，房间装饰奢华典雅。
5. 顶级套房的床品细部。

Index 索引

1. The Fairmont Palliser
Calgary, Canada
palliserhotel@fairmont.com
W.S. Maxwell

2. Beau-Rivage Palace
Lausanne, Switzerland
info@brp.ch
Jost, Bezencenet & Schnel

3. Taleon Imperial Hotel
Taleon Imperial Hotel
k.andreeva@taleon.ru
Jean-Baptiste Michel Vallin de la Mothe

4. Eynsham Hall
Oxford, UK
n.boguslawska@projectorange.com
Project Orange

5. Grand Hotel Kronenhof
St. Moritz, Switzerland
info@kronenhof.com
Justus Dahinden, Rolf Som

6. Villa Le Rose
Florence, Italy
pamela@villalerose.com
Leonardo and Beatrice Ferragamo

7. The Dorchester
London, UK
information.tdl@dorchestercollection.com
Champalimaud

8. Chateau Mcely
Mcely, The Czech Republic
karla@chateaumcely.com
Mr. Otto Bláha, Mrs Inez Cusmano

9. The Langham Huntington Hotel
Pasadena, USA
tllax.info@langhamhotels.com
The Johnson Studio

10. Grand Hotel Vesuvio
Naples, Italy
barbara.mancini@prestigehotels.it
Architect Sergio Bizzarro

11. Fairmont Grand Hotel Kyiv
Kyiv, Ukraine
kyiv@fairmont.com
Duangrit Bunnag

12. Bovey Castle
Exeter, UK
Katharine.walsh@delancey.com
Annabel Elliot

13. Charleston Place
Charleston, USA
executiveoffice@oeh.com
Cannon Design

14. The Milestone Hotel And Apartments
London, UK
akendall@rchmail.com
Bea Tollman

15. Hotel Beau Rivage, Geneve
Geneva, Switzerland
secom@beau-rivage.ch
Ms. Leila Corbett

16. Brenners Park-Hotel & Spa
Freiburg, Germany
information@brenners.com
Countess Bergit Douglas

17. Fairmont Le Montreux Palace
Montreux, Switzerland
joanna@impactasia.com
Fiona Thomson

图书在版编目（CIP）数据

新古典酒店艺术 / 任绍辉编；于芳，常文心译. -- 沈阳：辽宁科学技术出版社，2016.3
ISBN 978-7-5381-8918-6

Ⅰ. ①新… Ⅱ. ①任… ②于… ③常… Ⅲ. ①饭店－建筑设计 Ⅳ. ①TU247.4

中国版本图书馆CIP数据核字（2014）第269099号

出版发行：辽宁科学技术出版社
　　　　　（地址：沈阳市和平区十一纬路29号　邮编：110003）
印　刷　者：利丰雅高印刷（深圳）有限公司
经　销　者：各地新华书店
幅面尺寸：230 mm x290 mm
印　　张：14.5
字　　数：50千字
出版时间：2016年3月第1版
印刷时间：2016年3月第1次印刷
责任编辑：马竹音
封面设计：周　洁
封面设计：周　洁
责任校对：周　文

书号：ISBN 978-7-5381-8918-6
定价：258.00元

联系电话：024-23284360
邮购热线：024-23284502
http://www.lnkj.com.cn